SpringerBriefs in Electrical and Computer Engineering

More information about this series at http://www.springer.com/series/10059

Xiaowen Gong • Xu Chen • Lei Yang
Junshan Zhang

Social Group Utility Maximization

Springer

Xiaowen Gong
Arizona State University
Tempe
Arizona
USA

Xu Chen
University of Göttingen
Göttingen
Germany

Lei Yang
Arizona State University
Tempe
Arizona
USA

Junshan Zhang
Arizona State University
Tempe
Arizona
USA

ISSN 2191-8112 ISSN 2191-8120 (electronic)
ISBN 978-3-319-12321-9 ISBN 978-3-319-12322-6 (eBook)
DOI 10.1007/978-3-319-12322-6
Springer Cham Heidelberg New York Dordrecht London

Library of Congress Control Number: 2014954661

Printed on acid-free paper

Springer is part of Springer Science+Business Media (www.springer.com)

Preface

Cooperative networking has been widely recognized as a promising paradigm for a variety of wireless networking applications. However, little attention has been paid to exploiting the social network structure among wireless users to stimulate user cooperation in a systematic way. In this brief, we present a general social group utility maximization (SGUM) framework, which leverages the existing human social ties among wireless users to stimulate their cooperative behaviors in wireless networks. The SGUM framework advocates the notion of social group utility, which captures the diverse social ties among wireless users and the diverse physical relationships among their wireless devices in a unified manner. A notable merit of the SGUM framework is that it provides rich modeling flexibility and spans the continuum between non-cooperative game and network utility maximization—two traditionally disjoint paradigms for network optimization. To illustrate how to apply the SGUM framework for wireless networking, we study its application in several specific contexts.

In Chap. 1, we give an overview of mobile social networks and cooperative wireless networking. In Chap. 2, we formulate the SGUM framework, which comprises a social network graph model, a physical network graph model, and a SGUM game. In Chap. 3, we study the SGUM-based random access control and power control. For the SGUM-based random access control game and the SGUM-based power control game, we show that there exists a socially-aware Nash equilibrium (SNE). We also investigate the impact of social ties on users' strategies and social welfare. In Chap. 4, we study the SGUM-based database assisted spectrum access. We show that the SGUM-based spectrum access game is a potential game and thus always admits an SNE. Then we design a distributed spectrum access algorithm that can achieve an SNE with desirable social welfare. In Chap. 5, we study the SGUM-based pseudonym change for personalized location privacy. For the SGUM-based pseudonym change game, we show that there exists a SNE. Then we develop an algorithm that can efficiently find a Pareto-optimal SNE with desirable social welfare. In Chap. 5, we summarize the brief and discuss a generalization of the SGUM framework for future work.

We would like to thank our INEL colleagues for their valuable comments on the brief. We also would like to thank the Springer editors and staff for their great assistance in publishing this brief. This research work was supported in part by the U.S. National Science Foundation under Grants CNS-1117462, CNS-1218484, DoD MURI project No. FA9550-09-1-0643, and DTRA grant HDTRA1-13-1-0029.

Tempe, AZ, USA Xiaowen Gong
Goettingen, Germany Xu Chen
Tempe, AZ, USA Lei Yang
Tempe, AZ, USA Junshan Zhang

Contents

Chapter 1
Introduction

1.1 Mobile Social Networks

Mobile networks have been growing rapidly in the past few years and this trend will continue in the foreseeable future. Indeed, mobile phone shipments are projected to reach 1.9 billion units in 2014, which is about 7 times that of desktop and laptop combined [1]. Mobile data traffic is predicted to increase by over 100 times in the next ten years [2]. The widespread popularity of mobile networks has been driven by continuing advances of technologies. On one hand, advanced wireless communication technologies (e.g., MIMO, OFDM) have drastically improved the communication efficiency in existing wireless networks (e.g., cellular networks, WLANs). On the other hand, advanced mobile devices (e.g. smartphones) equipped with powerful sensors (e.g., cameras) and high computing capability have enabled a wide range of applications on mobile platforms. As a result, mobile networks have nowadays become an indispensable infrastructure in people's everyday life.

Different from other networks (e.g., sensor networks), a distinctive characteristic of mobile networks is that mobile devices are carried and operated by human beings. As a result, mobile users' interactions hinge heavily on human behavior. It is then natural to ask "How would mobile users' *social ties* influence their behaviors in mobile networks?" Social ties are built upon human social relationships (e.g., kinship, friendship, colleague relationship). Indeed, social ties play an unprecedented role in people's interactions with each other, mainly due to the explosive growth of online social networking services (e.g., Facebook, Twitter) in the past few years. In 2013, the number of online social network users worldwide has crossed 1.73 billion, nearly one quarter of the world's population [3]. With pervasive connectivity to the Internet via mobile devices, mobile users can interact with each other much more readily than ever before via online social networking services.

The social aspect of mobile networking is an emerging paradigm for network design and optimization. A survey of mobile social networking can be found in [5]. There has been some work using the social aspect of mobile users to enable user cooperation (e.g. cooperative forwarding [4, 6], cooperative relaying [7]). However, most of them do not consider that a user's cooperative behavior can affect multiple users, and affect different users to different extents. Furthermore, as a user's social

X. Gong et al., *Social Group Utility Maximization*, SpringerBriefs in Electrical and Computer Engineering, DOI 10.1007/978-3-319-12322-6_1

relationships with other users are generally *heterogeneous*, it would take into account the effect of its behavior on different users to different extents, which is not captured in the existing studies.

1.2 Cooperative Wireless Networking

Node cooperation has been widely recognized as a promising strategy for a variety of wireless networks. Indeed, individual nodes can achieve significant performance gain by cooperating in a coordinated way. For example, *cooperative communication* is an effective approach for improving the transmission rates among nodes in a communication network. In a cognitive radio network, *cooperative sensing* can enable cognitive radio (CR) users to efficiently detect spectrum opportunities that are not used by primary users (PUs). Although the benefit of node cooperation is pronounced, cooperative behaviors come at the cost of the cooperative nodes (e.g., in terms of the resource consumption devoted to cooperation). Therefore, for a network consisting of autonomous users, users may not be willing to cooperate without adequate incentives.

There exist numerous studies on incentive design for stimulating user cooperation for networking. Existing work on this subject can be broadly classified into three categories. One category of work makes use of reciprocity (also known as *barter*) [8–11]. Although a reciprocity-based approach is simple to implement, it is inefficient in general since it is rare to have synchronously matched requests for cooperation. Another category is based on (virtual) currency [12–15], in which a user earns currency by providing service to others and spends currency to receive service from others. The use of currency as a medium of exchange overcomes the shortcoming of reciprocity-based approaches by enabling users to "asynchronously trade" cooperation. However, a major drawback of using currency is that it incurs a significant implementation overhead, mainly due to the need to inhibit malicious manipulation among users without mutual trust. Consider, for example, the *Bitcoin* [16] that has recently drawn widespread attention as a digital currency. The creation and transfer of bitcoins need to consume considerable computing resources so that they can be secured against potential cheating using cryptographic tools. Reputation-based approaches [17–19] constitute the third category. Since reputation score can be viewed as a form of currency, these approaches share the same advantages and disadvantages as the currency-based ones.

For a network consisting of autonomous users (nodes) (e.g., ad hoc networks), each user may act in a selfish manner, in the sense that it only cares about its own benefit (e.g., utility) and does not care about the effect of its behavior on other users. In this case, the strategic interactions among users can be modeled by a *non-cooperative game* (NCG), where each user aims to maximize its payoff. NCG has been extensively studied for wireless networking applications [20]. Due to the selfish nature of users, the stable outcome of a non-cooperative game (e.g., a Nash equilibrium) may achieve a low social welfare (i.e., the total benefit of all users).

In contrast to selfish users, for a network where nodes are controlled by a central authority (e.g., sensor networks), all nodes are fully cooperative and aim to achieve the same system-wide goal. In this case, a widely used objective is *network utility maximization* (NUM), which is to maximize the total utility of all nodes. NUM has been widely studied for resource allocation in wireless networks [21].

Although there exists a significant body of work on NCG and NUM, very little attention has been paid to the continuum between these two extreme paradigms, especially in the context of mobile social networking. Recent work [22, 23] have studied the impact of altruistic behavior in a routing game. [24] has recently investigated a random access game between two symmetrically altruistic players.

References

1. Gartner: Worldwide PC, tablet and mobile phone shipments to grow 4.5 percent in 2013 (2013), http://www.gartner.com/newsroom/id/2610015
2. Cisco White paper: Global mobile data traffic data forecast update (2013–2018). (2014)
3. eMarketer: Social networking reaches nearly one in four around the world (2013), http://www.emarketer.com/Article/Social-Networking-Reaches-Nearly-One-Four-Around-World/1009976
4. P. Costa, C. Mascolo, M. Musolesi, G.P. Picco, Socially-aware routing for publish-subscribe in delay-tolerant mobile ad hoc networks. IEEE JSAC **26**(5) 748–760 (2008)
5. N. Kayastha, D. Niyato, P. Wang, E. Hossain, Applications, architectures, and protocol design issues for mobile social networks: A survey. Proc. IEEE **99**(12), 1439–1450 (2011)
6. W. Gao, Q. Li, B. Zhao, G. Cao, Multicasting in delay tolerant networks: A social network perspective. ACM MOBIHOC (2009)
7. X. Chen, B. Proulx, X. Gong, J. Zhang, Social trust and social reci-procity based cooperative D2D communications. ACM MOBIHOC (2013)
8. V. Srinivasan, P. Nuggehalli, C.-F. Chiasserini, R.R. Rao, Cooperation in wireless ad hoc networks. IEEE INFOCOM (2003)
9. W.H. Yuen, R.D. Yates, S.-C. Mau, Exploiting data diversity and multiuser diversity in non-cooperative mobile infostation networks. IEEE INFOCOM (2003)
10. P. Ganesan, M. Seshadri, On cooperative content distribution and the price of barter. IEEE ICDCS (2005)
11. U. Shevade, H.H. Song, L. Qiu, Y. Zhang, Incentive-aware routing in DTNs. IEEE ICNP (2008)
12. S. Zhong, J. Chen, Y.R. Yang, Incentive-aware routing in DTNs. IEEE INFOCOM (2003)
13. W. Wang, S. Eidenbenz, Y. Wang, X.-Y. Li, OURS: Optimal unicast routing systems in non-cooperative wireless networks. ACM MOBICOM (2006)
14. S. Zhong, F. Wu, On designing collusion-resistant routing schemes for non-cooperative wireless ad hoc networks. ACM MOBICOM (2007)
15. C. Zhang, X. Zhu, Y. Song, Y. Fang, C4: A new paradigm for providing incentives in multi-hop wireless networks. IEEE INFOCOM (2011)
16. Bitcoin: An open source P2P digital currency (2014), http://bitcoin.org/en/
17. S. Marti, T.J. Giuli, K. Lai, M. Baker, Mitigating routing misbehavior in mobile ad hoc networks. ACM MOBICOM (2000)
18. S. Buchegger, J.-Y.L. Boudec, Performance analysis of the CONFIDANT protocol. ACM MOBIHOC (2002)
19. J. Jaramillo, R. Srikant, DARWIN: Distributed and adaptive reputation mechanism for wireless networks. ACM MOBICOM (2007)
20. E. Altman, T. Boulogne, R. El-Azouzi, T. Jimenez, L. Wynter, A survey on networking games in telecommunications. Comput. Oper. Res. **33**, 286–311 (2006)

21. D. Palomar, M. Chiang, A tutorial to decompositon methods for network utility maximization. J. Sel. Areas Commun. **24**(8), 1439–1450 (2006)
22. P.-A. Chen, D. Kempe, Altruism, selfishness, and spite in traffic routing. ACM EC (2008)
23. M. Hoefer, A. Skopalik, Altruism in atomic congestion games. European Symposium on Algorithms (2009)
24. G. Kesidis, Y. Jin, A. Azad, E. Altman, Stable nash equilibria of ALOHA medium access games under symmetric, socially altruistic behavior. IEEE CDC (2010)

Chapter 2
Social Group Utility Maximization Framework

2.1 Motivation

As discussed in Chap. 1, the social ties among wireless users significantly influence their interactions with each other in wireless networks. One fundamental aspect of the positive social tie between two users is that they are altruistic to each other such that one cares about the other's welfare. As a result, a user would take into account the effect of its behavior on those having social ties[1] with it. It is then natural to ask "Is it possible to exploit users' social ties to stimulate their cooperative behaviors?" Indeed, altruistic behaviors are often observed among people with social ties. With this motivation, we view a wireless network as an overlay/underlay system (as illustrated in Fig. 2.1), where a "virtual social network" (social domain) overlays a physical communication network (physical domain). Wireless users are connected by social ties in the social domain, while their wireless devices are subject to physical relationships in the physical domain. It is important to observe that users generally have *diverse* social ties such that a user cares about others at different levels. For example, a user may care about her family members more than her friends, and cares about her friends more than an acquaintance of her. Similarly, it is clear that wireless devices also generally have *diverse* physical relationships. For example, depending on their physical locations, wireless devices can cause different levels of interference to each other. A primary goal here is to leverage the intrinsic diverse social tie structure among wireless users, which can be viewed as "hidden incentives" based on existing human relationships, to facilitate cooperative networking among their wireless devices subject to diverse physical relationships.

To this end, we advocate a social group utility maximization (SGUM) framework that takes into account both the diverse social coupling and diverse physical coupling among users. Specifically, we model the social coupling and physical coupling by a social graph and physical graph, respectively, and then we cast the distributed decision making problem among users as a SGUM game.

[1] In this brief, we use "social tie" to refer to "positive social tie" for brevity, and we will discuss "negative social tie" in Chap. 6.

X. Gong et al., *Social Group Utility Maximization*, SpringerBriefs in Electrical and Computer Engineering, DOI 10.1007/978-3-319-12322-6_2

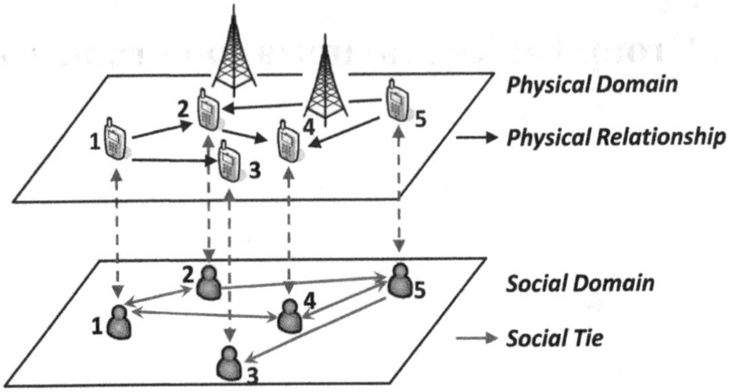

Fig. 2.1 Illustration of the social group utility maximization (SGUM) framework

2.2 Physical Network Graph Model

We consider a set of wireless users $\mathcal{N} = \{1, 2, \ldots, N\}$ where N is the total number of users. We denote the set of feasible strategies for each user $n \in \mathcal{N}$ as $\mathcal{X}_n$. For instance, a strategy $x \in \mathcal{X}_n$ can be choosing either a channel or a power level for wireless transmission. Subject to heterogeneous physical constraints, the strategy set $\mathcal{X}_n$ can be user-specific. For example, the strategy set $\mathcal{X}_n$ can be a set of feasible relay users that are in vicinity of user n for cooperative communication.

To capture the diverse physical coupling among the users in the physical domain, we introduce a *physical graph* $\mathcal{G}^p = \{\mathcal{N}, \mathcal{E}^p\}$ (see Fig. 2.1 for an example). Here the set of users $\mathcal{N}$ is the vertex set, and $\mathcal{E}^p \equiv \{(n,m) : e^p_{nm} = 1, \forall n, m \in \mathcal{N}\}$ is the edge set where $e^p_{nm} = 1$ if and only if users n and m have physical coupling (e.g., cause interference to each other). We also denote the set of users that have physical coupling with user n as $\mathcal{N}^p_n \equiv \{m \in \mathcal{N} : e^p_{nm} = 1\}$.

Let $\boldsymbol{x} = (x_1, \ldots, x_N) \in \prod_{n=1}^{N} \mathcal{X}_n$ be the strategy profile of all users. Given the strategy profile $\boldsymbol{x}$, the individual utility function of user n is denoted as $u_n(\boldsymbol{x})$, which represents the payoff of user n, accounting for the physical coupling among users. For example, $u_n(\boldsymbol{x})$ can be the achieved data rate or the satisfaction of quality of service (QoS) requirement of user n under the strategy profile $\boldsymbol{x}$. Note that in general the underlying physical graph plays a critical role in determining the individual utility $u_n(\boldsymbol{x})$. For example, users' achieved data rates are determined by the interference graph and channel quality.

2.3 Social Network Graph Model

To capture the diverse social coupling among the users in the social domain, we introduce a *social graph* $\mathcal{G}^s = \{\mathcal{N}, \mathcal{E}^s\}$ to model their social ties. Here the edge set is given by $\mathcal{E}^s = \{(n,m) : e^s_{nm} = 1, \forall n, m \in \mathcal{N}\}$ where $e^s_{nm} = 1$ if and only if users

n has a social tie with user m, which can be built on, e.g., the kinship, friendship, or colleague relationship between them. We denote the *social tie level* from user n to user m as s_{nm}. We assume that each user n's social tie level to itself is $s_{nn} = 1$, and we normalize user n's social tie level to user $m \neq i$ as $s_{nm} \in (0, 1]$, which represents the extent to which user n cares about user m relative to user n cares about itself, with a greater value of s_{nm} indicating a stronger social tie. We also assume that $s_{nm} = 0$ if no social tie exists from user n to user m. We define user n's *social group* $\mathcal{N}_n^s$ as the set of users that have social ties with user n, i.e., $\mathcal{N}_n^s = \{m : e_{nm}^s = 1, \forall m \in \mathcal{N}\}$.

Based on the physical and social graph models described above, users are coupled in the physical domain due to the physical relationships, and are also coupled in the social domain due to their social ties. With this insight, we define the *social group utility* of each user n as

$$f_n(\boldsymbol{x}) = u_n(\boldsymbol{x}) + \sum_{m \in \mathcal{N}_n^s} s_{nm} u_m(\boldsymbol{x}). \tag{2.1}$$

It follows that the social group utility of each user consists of two parts: (1) its own individual utility and (2) the weighted sum of the individual utilities of other users having social tie with it. In a nutshell, the social group utility function captures that each user is socially-aware and cares about the users having social tie with it.

2.4 Social Group Utility Maximization Game

We consider the distributed decision making problem among the users for maximizing their social group utilities. Let $\boldsymbol{x}_{-n} = (x_1, \ldots, x_{n-1}, x_{n+1}, \ldots, x_N)$ be the set of strategies chosen by all other users except user n. Given the other users' strategies $\boldsymbol{x}_{-n}$, user n aims to choose a strategy $x_n \in \mathcal{X}_n$ thatmaximizes its social group utility, i.e.,

$$\max_{x_n \in \mathcal{X}_n} f_n(x_n, \boldsymbol{x}_{-n}), \forall n \in \mathcal{N}.$$

The distributed nature of the problem above naturally leads to a formulation based on game theory such that each user aims to maximize its social group utility. We thus formulate the decision making problem among the users as a strategic game $\Gamma = (\mathcal{N}, \{\mathcal{X}_n\}_{n \in \mathcal{N}}, \{f_n\}_{n \in \mathcal{N}})$, where the set of users $\mathcal{N}$ is the set of players, $\mathcal{X}_n$ is the set of strategies for each user n, and the social group utility function f_n of each user n is the payoff function of player n. In the sequel, we call the game Γ as the SGUM game. We next introduce the concept of *socially-aware Nash equilibrium* (SNE).

Definition 2.1 A strategy profile $\boldsymbol{x}^* = (x_1^*, \ldots, x_N^*)$ is a socially-aware Nash equilibrium of the SGUM game if no player can improve its social group utility by unilaterally changing its strategy, i.e.,

$$x_n^* = \arg\max_{x_n \in \mathcal{X}_n} f_n(x_n, \boldsymbol{x}_{-n}), \forall n \in \mathcal{N}.$$

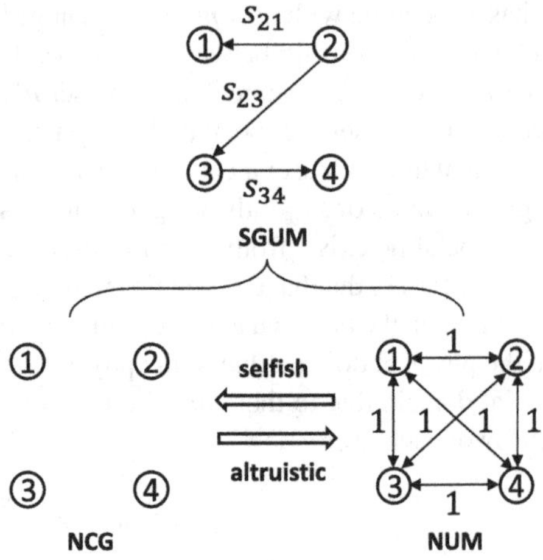

Fig. 2.2 The social group utility maximization (SGUM) game captures non-cooperative game (NCG) and network utility maximization (NUM) as special cases

It is worth noting that under different social graphs, the proposed SGUM game formulation can provide rich flexibility for modeling network optimization problems (as illustrated in Fig. 2.2). When the social graph consists of isolated nodes with $s_{nm} = 0$ for any $n, m \in \mathcal{N}$ (i.e., all users are socially-oblivious), the SGUM game degenerates to a standard non-cooperative game. When the social graph is fully meshed with edge weight $s_{nm} = 1$ for any $n, m \in \mathcal{N}$ (i.e., all users are fully altruistic), the SGUM game becomes a network utility maximization problem, which aims to maximize the system-wide utility. The SGUM framework can be applied with general social graphs and thus can bridge the gap between non-cooperative game and network utility maximization—two traditionally disjoint paradigms for network optimization (as illustrated in Fig. 2.3). These two paradigms are captured under the SGUM framework as two special cases where no social tie exists among users, and all users are connected by strongest social ties, respectively.

We emphasize that the SGUM game is quite different from a *coalitional game* [1], since each user in the latter aims to maximize its individual benefit (although it is achieved by cooperating with other users). Furthermore, while each user in a coalitional game can only participate in *one* coalition, a user in the SGUM game can be in *multiple* social groups of different users.

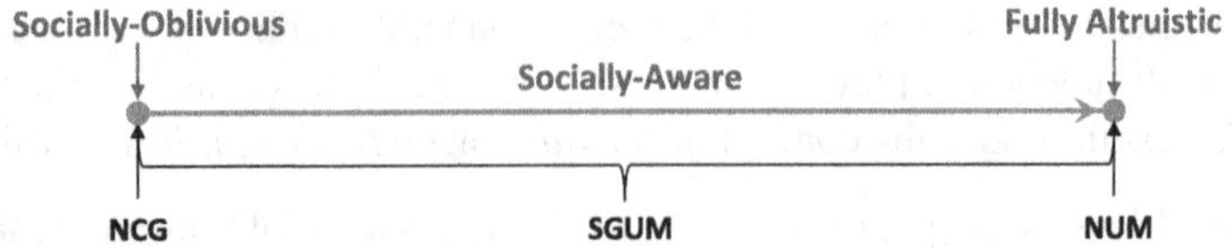

Fig. 2.3 The social group utility maximization (SGUM) framework spans the continuum between non-cooperative game (NCG) and network utility maximization (NUM)

The SGUM is a general framework that can be applied for a wide range of wireless networking applications. To get a more concrete sense of the framework, in the rest of this brief, we will study its application in a number of specific contexts.

Reference

1. M. Osborne, A. Rubinstein, A Course in Game Theory (MIT Press, Cambridge, 1994)

Chapter 3
SGUM-based Random Access Control and Power Control

In this chapter, we study the application of the SGUM framework to two classical problems for wireless networks: random access control and power control.

3.1 Introduction

Wireless spectrum is a limited resource shared by wireless users. Due to the broadcast nature of wireless communication, wireless nodes in physical proximity are subject to interference to each other if they transmit concurrently on the shared wireless spectrum. Multiple access methods have been developed to allow wireless nodes to share the use of wireless resources in an interference-free manner. In contrast to contention-free multiple access (e.g., TDMA, FDMA) which relies on centralized coordination, contention-based random access allows contending wireless nodes to share wireless spectrum in a distributed manner. On the other hand, interference-free environment may not be available in some wireless networks. For example, in CDMA systems, perfect orthogonality among users' transmit signals is difficult to achieve, and thus a user's transmission is affected by the interference power received from other users. In interference-limited wireless networks, power control represents a key degree of freedom for network design and optimization. Game theory has been extensively applied to study the strategic decision making among autonomous and rational users for both random access control and power control. A survey of random access control games and power control games can be found in [1] and [2], respectively. To stimulate user cooperation for efficient spectrum sharing, we cast random access control and power control among userswith social ties as SGUM games.

3.2 SGUM-based Random Access Control

3.2.1 System Model

We consider a set of users under the protocol interference model, where each user i is a link consisting of transmitter T_i and receiver R_i. For example, in Fig. 3.1, T_1 interferes with R_2, T_2 interferes with R_1, T_3 interferes with R_1, where dashed circles

X. Gong et al., *Social Group Utility Maximization*, SpringerBriefs in Electrical and Computer Engineering, DOI 10.1007/978-3-319-12322-6_3

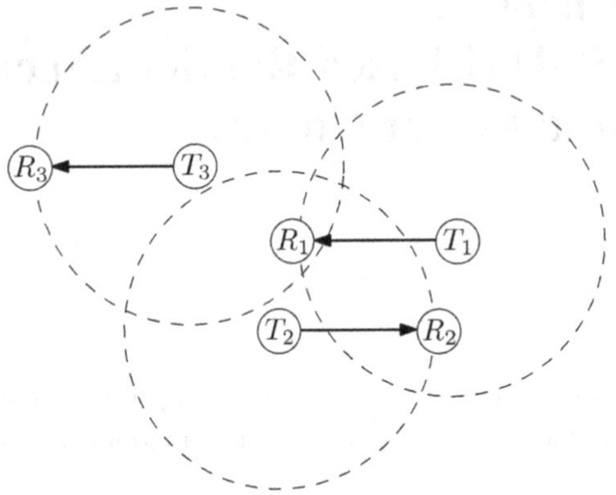

Fig. 3.1 An example of three links under the protocol interference model

define the interference ranges of transmitters. Let $\mathcal{I}_i^+$ denote the set of receivers that transmitter T_i causes interference to, and $\mathcal{I}_i^-$ denote the set of transmitters that causes interference to receiver R_i. In a time-slotted system, each user i contends for the opportunity of data transmission with probability $q_i \in [0, 1]$ in a time slot. If multiple interfering links contend in the same time slot, a collision occurs and no link can grab the transmission opportunity. Then the probability b_i that user i can grab the transmission opportunity is given by

$$b_i(q_i, \boldsymbol{q}_{-i}) = q_i \prod_{j \in \mathcal{I}_i^-} (1 - q_j). \tag{3.1}$$

We assume that the individual utility of user i is given by

$$u_i(q_i, \boldsymbol{q}_{-i}) = \log(\theta_i b_i) - c_i q_i \tag{3.2}$$

where $\theta_i > 0$ represents user i's efficiency of utilizing the transmission opportunity (e.g., transmission rate), and $c_i > 0$ represents user i's cost of contention. Note that the logarithmic function is widely used for modeling the utility of wireless users [3, 4]. Then, under the SGUM framework, we define the SGUM-based random access control game as $G \triangleq (\mathcal{N}, \{q_i\}, \{f_i\})$, where

$$f_i(q_i, \boldsymbol{q}_{-i}) = \log\left(\theta_i q_i \prod_{j \in \mathcal{I}_i^-} (1 - q_j)\right) - c_i q_i$$

$$\sum_{j \neq i} s_{ij} \left[\log\left(\theta_j q_j \prod_{k \in \mathcal{I}_j^-} (1 - p_k)\right) - c_j q_j\right]. \tag{3.3}$$

3.2.2 Game Analysis

We first have the following result.

Theorem 3.1 *For the SGUM-based random access control game, there exists a unique SNE, which is*

$$q_i^{SNE} = \frac{\sum_{j\in\mathcal{I}_i^+} s_{ij} + 1 + c_i - \sqrt{(\sum_{j\in\mathcal{I}_i^+} s_{ij} + 1 + c_i)^2 - 4c_i}}{2c_i}, \forall i \in \mathcal{N}. \quad (3.4)$$

Proof . Using (3.3), setting the first-order derivative of $f_i(q_i, \boldsymbol{q}_{-i})$ to 0, we have

$$\frac{\partial f_i(q_i, \boldsymbol{q}_{-i})}{\partial q_i} = \frac{1}{q_i} - \sum_{j\in\mathcal{I}_i^+} \frac{s_{ij}}{1-q_i} - c_i$$
$$= \frac{c_i q_i^2 - \left(\sum_{j\in\mathcal{I}_i^+} s_{ij} + 1 + c_i\right) q_i + 1}{q_i(1-q_i)} = 0. \quad (3.5)$$

Then we obtain the smaller root of Eq. (3.5) as

$$\frac{\sum_{j\in\mathcal{I}_i^+} s_{ij} + 1 + c_i - \sqrt{\left(\sum_{j\in\mathcal{I}_i^+} s_{ij} + 1 + c_i\right)^2 - 4c_i}}{2c_i}$$
$$\leq \frac{1 + c_i - \sqrt{(1+c_i)^2 - 4c_i}}{2c_i} \leq 1 \quad (3.6)$$

where the first inequality follows from that the first-order derivative of the small root with respect to s_{ij} is

$$\frac{1}{2c_i}\left(1 - \frac{\sum_{j\in\mathcal{I}_i^+} s_{ij} + 1 + c_i}{\sqrt{(\sum_{j\in\mathcal{I}_i^+} s_{ij} + 1 + c_i)^2 - 4c_i}}\right) < 0. \quad (3.7)$$

We also obtain the larger root of Eq. (3.5) as

$$\frac{\sum_{j\in\mathcal{I}_i^+} s_{ij} + 1 + c_i + \sqrt{\left(\sum_{j\in\mathcal{I}_i^+} s_{ij} + 1 + c_i\right)^2 - 4c_i}}{2c_i}$$
$$\geq \frac{1 + c_i + \sqrt{(1+c_i)^2 - 4c_i}}{2c_i} \geq 1.$$

Therefore, the SNE strategy q_i^{SNE} is unique and is the smaller root of Eq. (3.5). □

The result below directly follows from Theorem 3.1 and (3.7).

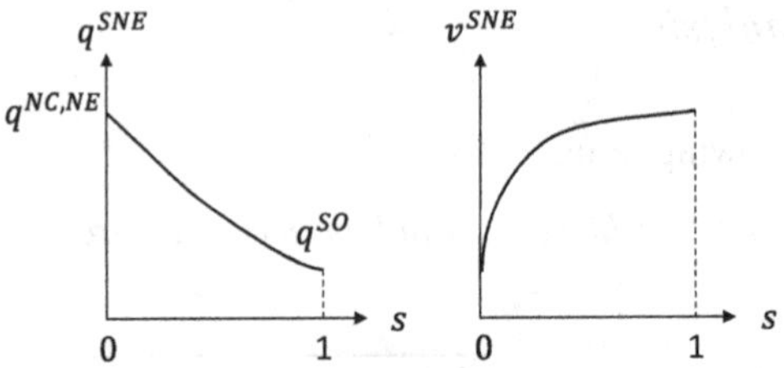

Fig. 3.2 For a two-user SGUM game for random access control, as the social tie level $s \triangleq s_{12} = s_{21}$ increases from 0 to 1, each user's SNE strategy q^{SNE} migrates from its NE strategy $q^{NC,NE}$ for a standard NCG to its social optimal strategy q^{SO} for NUM, and the social welfare v^{SNE} of the SNE also migrates correspondingly

Corollary 3.1 *Each user's access probability at the SNE is decreasing as its social tie levels with others increase.*

Remark 3.1. We observe that each user's SNE strategy does not depend on other users' strategies (also known as a *dominant strategy*), but depends on the user's social ties with others. Clearly, when a user increases its access probability, it also increases the collision probabilities of the users within its interference range, and thus reduces their individual utilities. Therefore, a user would decrease its access probability when its social ties with those within its interferencerange get stronger (as illustrated in Fig. 3.2).

Let $V(\boldsymbol{q})$ denote the social welfare of all users, i.e., the total individual utility of all users:

$$V(\boldsymbol{q}) \triangleq \sum_{i=1}^{N} \left[\log \left(\theta_i q_i \prod_{j \in \mathcal{I}_i^-} (1 - q_j) \right) - c_i q_i \right]. \tag{3.8}$$

Proposition 3.1 *The social welfare of the SNE is increasing as social tie levels increase, and reaches the social optimal point when all socialtie levels are equal to 1.*

Proof. Using (3.8), setting the first-order derivative of $V(\boldsymbol{q})$ to 0, we have

$$\frac{\partial V(\boldsymbol{q})}{\partial q_i} = \frac{c_i q_i^2 - (|\mathcal{I}_i^+| + 1 + c_i) q_i + 1}{q_i(1 - q_i)} = 0. \tag{3.9}$$

Similar to the proof of Theorem 3.1, we obtain the social optimal strategy q_i^{SO} that maximizes $V(\boldsymbol{q})$ as the smaller root of Eq. (3.9), which is

$$q_i^{SO} = \frac{|\mathcal{I}_i^+| + 1 + c_i - \sqrt{\left(|\mathcal{I}_i^+| + 1 + c_i\right)^2 - 4c_i}}{2c_i}.$$

Since the larger root of Eq. (3.9) is

$$\frac{|\mathcal{I}_i^+| + 1 + c_i + \sqrt{(|\mathcal{I}_i^+| + 1 + c_i)^2 - 4c_i}}{2c_i}$$

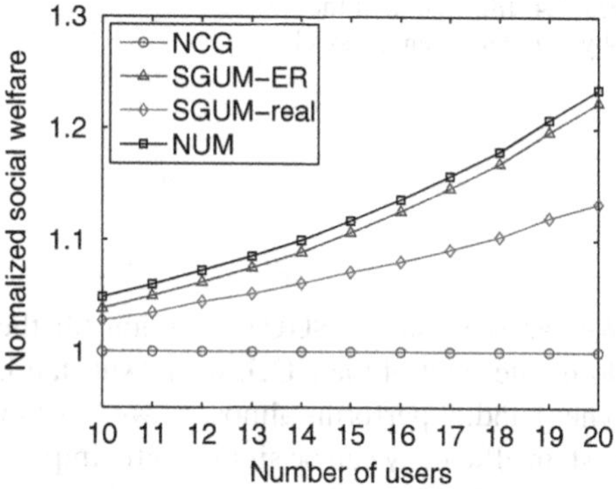

Fig. 3.3 Impact of number of users

$$\geq \frac{1 + c_i + \sqrt{(1 + c_i)^2 - 4c_i}}{2c_i} \geq 1,$$

we have $\frac{\partial V(\boldsymbol{q})}{\partial q_i} < 0$ for $q_i \in [q_i^{SO}, 1]$, and thus $V(\boldsymbol{q})$ is decreasing in q_i when $q_i \in [q_i^{SO}, 1]$. Using Corollary 3.1, q_i^{SNE} is decreasing in $s_{ij}, \forall j \in \mathcal{I}_i^+, \forall i \in \mathcal{N}$, and hence $V(\boldsymbol{q}^{SNE})$ is increasing in $s_{ij}, \forall j \in \mathcal{I}_i^+, \forall i \in \mathcal{N}$. □

Remark 3.2 . Intuitively, since the social welfare is equal to users' individual utilities summed up with the same weight 1, a user's SNE strategy is closer to the social optimal strategy when other users weigh more in that user's social group utility (i.e., the social tie levels to them increase), and the social welfare increases. As social tie levels increase, a user's SNE strategy migrates from its NE strategy for a standard NCG to its social optimal strategy for NUM (as illustrated in Fig. 3.2). This demonstrates that the SGUM game framework spans the continuum between these traditionally disjoint paradigms.

3.2.3 Numerical Results

We consider N users each of which is a link consisting of a transmitter and a receiver. Each transmitter or receiver is randomly located in a square area with side length 500 m. Under the protocol interference model, we assume that a link causes interference to another link if the former link's transmitter is within 100 m of the latter link's receiver. We simulate the social graph based on both the Erdos-Renyi (ER) model with link probability 0.5 and the real data trace of the friendship network Brightkite. We assume that the strength of a social tie is 1 if the social tie exists.

To illustrate the system efficiency of the SGUM solution, we compare it with the NCG solution where each user aims to maximize its individual utility, and the NUM solution where the total individual utility of all users is maximized. Figure 3.3 depicts the social welfare of the SNE for SGUM and the social optimal solution for NUM normalized with respect to the NE for NCG, as the number of users increases.

Fig. 3.4 Illustration of the physical interference model

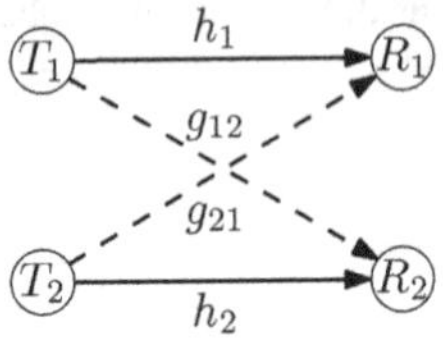

We can see that the SGUM solution for the ER model based social graph always dominates that of the NCG, with a substantial performance gain up to 22 %. On the other hand, it performs almost as well as the NUM solution. This demonstrates that system efficiency can be significantly improved by exploiting social ties. We observe that the SGUM solution for the real data based social graph is worse than that for the ER model based social graph due to that social ties are weaker in the real data than in the ER graph with link probability 0.5. However, it still can achieve a performance gain up to 13 % over that of the NCG solution.

3.3 SGUM-based Power Control

3.3.1 *System Model*

We consider a set of users under the physical interference model, where each user i is a link consisting of a transmitter T_i and a receiver R_i. The channel gain of communication link i is h_i, and the channel gain of the interference link between transmitter T_i and receiver R_j is g_{ij} (as illustrated in Fig. 3.4). The noise at receiver R_i is n_i. Then the signal tointerference and noise ratio (SINR) γ_i of link i is given by

$$\gamma_i(p_i, \boldsymbol{p}_{-i}) = \frac{h_i p_i}{n_i + \sum_{j=1}^{N} g_{ji} p_j}$$

where p_i denotes the transmit power of T_i. We assume that the individual utility u_i of player i is given by

$$u_i(p_i, \boldsymbol{p}_{-i}) = \log(\gamma_j) - c_i p_i$$

where c_i denotes the cost of per unit power consumption. Similar to Sect. 3.2, we also use the logarithmic function to model the utility of a user. For example, $\log(\gamma_i)$ can be a good approximation for the channel capacity $\log(1 + \gamma_i)$ under the high SINR regime. Also, $\log(\gamma_i)$ can be used to quantify the satisfaction of wireless users' requirements in terms of SINR. Then, under the SGUM framework, we define the SGUM game for power control as $G \triangleq (\mathcal{N}, \{p_i\}, \{f_i\})$, where

$$f_i(p_i, \boldsymbol{p}_{-i}) = \log\left(\frac{h_i p_i}{n_i + \sum_{j \neq i} g_{ji} p_j}\right) - c_i p_i +$$

$$\sum_{k\neq i} s_{ik}\left(\log\left(\frac{h_k p_k}{n_k+\sum_{j\neq k} g_{jk} p_j}\right)-c_k p_k\right). \tag{3.10}$$

3.3.2 *Game Analysis*

We first have the following result.

Theorem 3.2 *The SGUM-based power control game is a supermodular game, and thus there exists at least one SNE.*

Proof. Using (3.10), we have

$$\frac{\partial f_i(p_i,\boldsymbol{p}_{-i})}{\partial p_i}=\frac{1}{p_i}-\sum_{k\neq i}\frac{s_{ik}g_{ik}}{n_k+\sum_{j\neq k} g_{jk} p_j}-c_i.$$

Since each term in theabove summation term is decreasing in p_j, $\forall j \in \mathcal{N}\setminus i$, it follows that

$$\frac{\partial^2 f_i(p_i,\boldsymbol{p}_{-i})}{\partial p_i \partial p_j}>0, \forall j\in\mathcal{N}\setminus i$$

which implies that the social group utility function $f_i(p_i,\boldsymbol{p}_{-i})$ is supermodular. It follows from [5] that there exists at least one NE. □

Since the SGUM-based power control game is a supermodular game, it follows from [6] that users can start from any strategies (e.g., $\boldsymbol{p}=(0,\cdots,0)$) and use asynchronous *best response* updates such that their strategies will monotonically converge to a SNE.

For ease of exposition, in the rest of this section we will focus on the SGUM-based power control game with *two* users, because the two-user case can shed light on the impact of social ties on users' strategies and social welfare. Furthermore, in general, the game with more than two users does not yield closed-form SNE strategies, and hence is much more difficult to quantify the impact.

Theorem 3.3. *For the two-user SGUM-based power control game, there exists a unique SNE, which is*

$$p_1^{SNE}=\sqrt{\alpha_1^2+\beta_1}-\alpha_1,\ p_2^{SNE}=\sqrt{\alpha_2^2+\beta_2}-\alpha_2$$

where

$$\alpha_1\equiv\frac{s_{12}g_{12}+c_1n_2-g_{12}}{2c_1g_{12}},\ \beta_1\equiv\frac{n_2}{c_1g_{12}}$$

and

$$\alpha_2\equiv\frac{s_{21}g_{21}+c_2n_1-g_{21}}{2c_2g_{21}},\ \beta_2\equiv\frac{n_1}{c_2g_{21}}.$$

Proof.

$$u_1(p_1,p_2) = \log\left(\frac{h_1 p_1}{n_1 + g_{21}p_2}\right) - c_1 p_1 + s_{12}\log\left(\frac{h_2 p_2}{n_2 + g_{12}p_1}\right) - s_{12}c_2 p_2,$$

we have

$$\frac{\partial u_1(p_1,p_2)}{\partial p_1} = \frac{1}{p_1} - \frac{s_{12}g_{12}}{n_2 + g_{12}p_1} - c_1.$$

Since

$$\lim_{p_1\to 0}\left(\frac{1}{p_1} - \frac{s_{12}g_{12}}{n_2 + g_{12}p_1}\right) \geq \lim_{p_1\to 0}\left(\frac{1}{p_1} - \frac{s_{12}}{p_1}\right) = \infty$$

and

$$\lim_{p_1\to\infty}\left(\frac{1}{p_1} - \frac{s_{12}g_{12}}{n_2 + g_{12}p_1}\right) = 0$$

and

$$\begin{aligned}\frac{\partial\left(\frac{1}{p_1} - \frac{s_{12}g_{12}}{n_2+g_{12}p_1}\right)}{\partial p_1} &= -\frac{1}{p_1^2} + \frac{s_{12}g_{12}^2}{(n_2 + g_{12}p_1)^2} \\ &= \frac{(s_{12}-1)g_{12}^2p_1^2 - 2n_2g_{12}p_1 - n_2^2}{p_1^2(n_2 + g_{12}p_1)^2} \\ &< 0,\end{aligned}$$

there exists a unique value of p_1 such that

$$\frac{1}{p_1} - \frac{s_{12}g_{12}}{n_2 + g_{12}p_1} - c_1 = 0, \tag{3.11}$$

which is also the value of p_1^{SNE}. Solving (3.11), we obtain the desired result. Similarly, we can obtain p_2^{SNE}. □

Next we have the following result.

Corollary 3.2. *For the two-user SGUM-based power control game, each user's transmit power at the SNE is decreasing as its social tie level with the other increases.*

Proof.

$$p_1^{SNE} = \sqrt{\alpha_1^2 + \beta_1} - \alpha_1$$

and

$$\alpha_1 \equiv \frac{s_{12}g_{12} + c_1n_2 - g_{12}}{2c_1g_{12}},\ \beta_1 \equiv \frac{n_2}{c_1g_{12}} > 0,$$

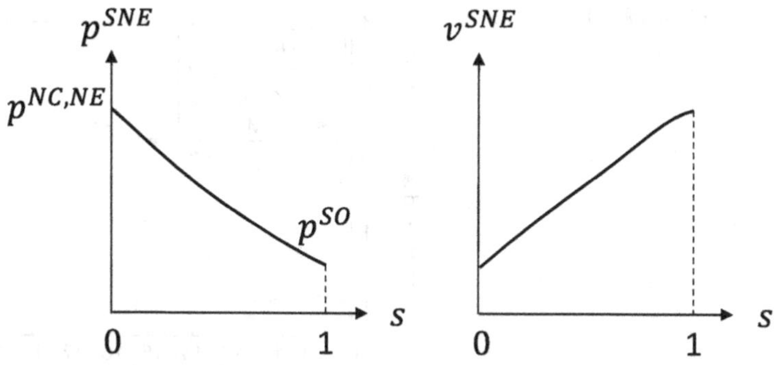

Fig. 3.5 For the two-user SGUM-based power control game, as the social tie level $s \triangleq s_{12} = s_{21}$ increases from 0 to 1, each user's SNE strategy p^{SNE} migrates from its NE strategy $p^{NC,NE}$ for a standard NCG to its social optimal strategy p^{SO} for NUM, and the social welfare v^{SNE} of the SNE also migrates correspondingly

we have

$$\frac{\partial p_1^{SNE}}{\partial s_{12}} = \frac{\partial \left(\sqrt{\alpha_1^2 + \beta_1} - \alpha_1\right)}{\partial \alpha_1} \frac{\partial \alpha_1}{\partial s_{12}}$$

$$= \left(\frac{\alpha_1}{\sqrt{\alpha_1^2 + \beta_1}} - 1\right) \frac{1}{2c_1} < 0.$$

So p_1^{SNE} is decreasing in s_{12}. Similarly, we can show that p_2^{SNE} is decreasing in s_{21}. □

Proposition 3.2 *For the two-user SGUM-based power control game, the social welfare of the SNE is increasing as social tie levels increase, and reaches the social optimal point when all social tie levels are equal to 1.*

Proof. Since

$$V(p_1, p_2) = \log\left(\frac{h_1 p_1}{n_1 + g_{21} p_2}\right) - c_1 p_1$$

$$+ \log\left(\frac{h_2 p_2}{n_2 + g_{12} p_1}\right) - c_2 p_2$$

we have

$$\frac{\partial V(p_1, p_2)}{\partial p_1} = \frac{1}{p_1} - \frac{g_{12}}{n_2 + g_{12} p_1} - c_1.$$

Using the same argument as in the proof of Theorem 3.3., the optimal value p_1^{SO} of p_1 for $V(p_1, p_2)$ is the unique solution of

$$\frac{1}{p_1} - \frac{g_{12}}{n_2 + g_{12} p_1} - c_1 = 0.$$

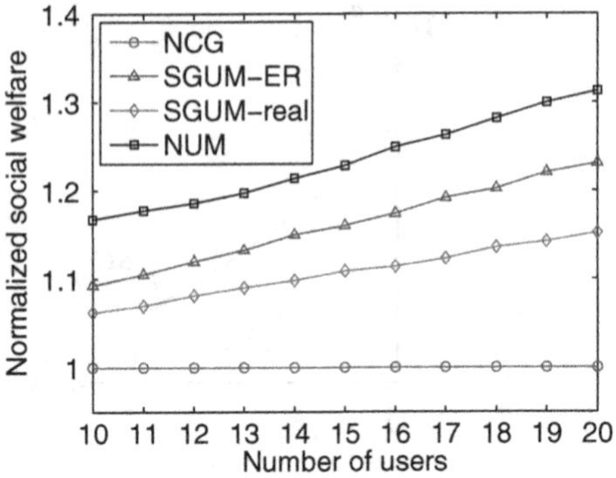

Fig. 3.6 Impact of number of users

In particular, we have $p_1^{SNE} \geq p_1^{SO}$. Since $\frac{\partial V(p_1,p_2)}{\partial p_1} < 0$ when $p_1 \geq p_1^{SO}$, $V(p_1, p_2)$ is decreasing in p_1 when $p_1 \geq p_1^{SO}$. Using Lemma 3.2., p_1^{SNE} is decreasing in s_{12}, and hence $V(p_1^{SNE}, p_2^{SNE})$ is increasing in s_{12} since p_2^{SNE} is independent of s_{12}. Similarly, we can show that $V(p_1^{SNE}, p_2^{SNE})$ is increasing in s_{21}. □

Remark 3.3. Similar to the SGUM-based random access control game, for the two-user SGUM-based power control game, each user's strategy at the SNE is also a dominant strategy. As a user's social tie level with the other increases, the user's transmit power at the SNE decreases, and the social welfare increases. Therefore, as the social tie level increases, a user's SNE strategy migrates from its NE strategy for a standard NCG to its social optimal strategy for NUM (as illustrated in Fig. 3.5).

3.3.3 Numerical Results

We consider N users each of which is a link consisting of a transmitter and a receiver. Each transmitter or receiver is randomly located in a square area with side length 500 m. Under the physical interference model, we assume that the channel condition of a link (communication or interference link) only depends on the path loss effect with path loss factor 3. We assume that the transmit power of each link is 1 W and the noise power at each receiver is 0.1 W.

Figure 3.6 shows the normalized social welfare for a varying number of users. We can see that the SGUM solution for the ER model based social graph can achieve a performance gain up to 23 % over the NCG solution, and its performance loss from the NUM solution is at most 10 %. The SGUM solution for the real data based social graph can achieve a performance gain up to 15 %.

3.4 Summary

In this chapter, we study the SGUM-based random access control and power control. For the SGUM-based random access control game, we derive the unique SNE. For the SGUM-based power control game, we show that it is a supermodular game and thus there exists an SNE. We also derive the unique SNE for the two-user case of the SGUM-based power control game. For both games, we show that as social tie levels increase, each user's SNE strategy is decreasing and the social welfare of the SNE is increasing. Our findings provide useful insights into the impact of social ties on users' strategies and social welfare.

References

1. K. Akkarajitsakul, E. Hossain, D. Niyato, D.I. Kim, Game theoretic approaches for multiple access in wireless networks: A survey. IEEE Commun. Surv. Tutor. **13**(3), 372–395, (2011)
2. M. Chiang, P. Handy, T. Lan, C.W. Tan, Power control in wireless cellular networks. Found. Trends Netw. **2**(4), 381–533, (2008)
3. G. Scutari, S. Barbarossa, D.P. Palomar, Potential games: A framework for vector power control problems with coupled constraints. IEEE ICASSP (2006)
4. U.O. Candogan, I. Menache, A. Ozdaglar, P.A. Parrilo, Near-optimal power control in wireless networks: A potential game approach. IEEE INFOCOM (2010)
5. D.M. Topkis, *Supermodularity and Complementarity* (Princeton University Press, 1998)
6. E. Altman, Z. Altman, S-modular games and power control in wireless networks. IEEE Trans. Autom. Control **48**, 839–842, (May 2003)

Chapter 4
SGUM-based Database Assisted Spectrum Access

In this chapter, we study the application of the SGUM framework to database assisted spectrum access.

4.1 Introduction

The very recent FCC ruling requires that white-space users (i.e., secondary TV spectrum users) must rely on a geo-location database to determine the spectrum availability [1]. Although the database-assisted approach obviates the need of spectrum sensing by individual users, it remains challenging to achieve reliable shared spectrum access, because different white-space users may choose to access the same vacant channel and thus incur severe interference to each other. To stimulate effective cooperation for channel allocation among white-space users, we cast the database assisted distributed spectrum access problem among white-space users with social ties as a SGUM game.

4.2 System Model

We consider a set of white-space users $\mathcal{N} = \{1, 2, \ldots, N\}$ where N is the total number of users. We denote the set of TV channels as $\mathcal{M} = \{1, 2, \ldots, M\}$. According to the recent ruling by FCC [1], to protect the incumbent primary TV users, each white-space user $n \in \mathcal{N}$ will first send a spectrum access request message containing its geo-location information to a Geo-location database (see Fig. 4.1 for an illustration). The database then feeds back the set of vacant channels $\mathcal{M}_n \in \mathcal{M}$ and the allowable transmission power level P_n to user n. The ruling by FCC indicates that the allowable transmission power limit for personal/portable white-space devices (e.g., mobile phones) is 100 mW [1]. For ease of exposition, we hence assume that each user n accesses the white-space spectrum with the same power level. Each user n then chooses a feasible channel a_n from the vacant channel set $\mathcal{M}_n$ for data transmission.

X. Gong et al., *Social Group Utility Maximization*, SpringerBriefs in Electrical and Computer Engineering, DOI 10.1007/978-3-319-12322-6_4

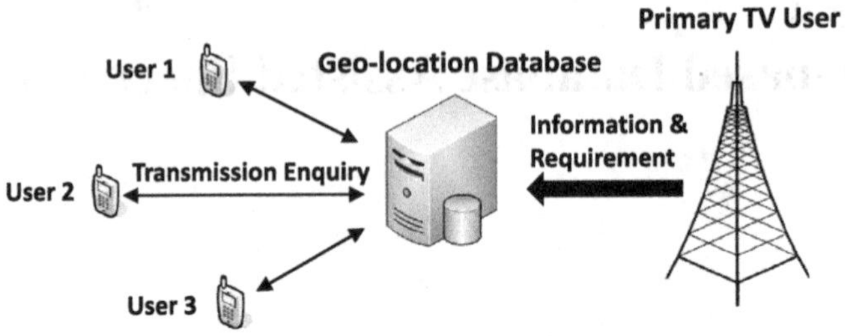

Fig. 4.1 An illustration of database assisted spectrum access

Although the database-assisted approach obviates the need of spectrum sensing by individual users, it remains challenging to achieve reliable distributed spectrum access, because many different white-space users may choose to access the same vacant channel and thus incur severe interference to each other [2, 3].

To stimulate effective cooperation among users for interference mitigation, we leverage the social ties among users and apply the SGUM approach. To capture the physical coupling, we construct the interference graph $\mathcal{G}^p = \{\mathcal{N}, \mathcal{E}^p\}$ based on the interference relationships among users. Here the set of white-space users $\mathcal{N}$ is the vertex set, and $\mathcal{E}^p \equiv \{(n,m) : e^p_{nm} = 1, \forall n, m \in \mathcal{N}\}$ is the edge set where $e^p_{nm} = 1$ if and only if users n and m can generate significant interference and affect the data transmissions of each other. For example, we can construct the interference graph $\mathcal{G}^p$ based on spatial relationships of the users [4]. Let δ denote the transmission range of each user. We then have $e^p_{nm} = 1$ if and only if the distance d_{nm} between user n and m is not greater than the threshold δ, i.e., $d_{nm} \leq \delta$.

Let $\boldsymbol{a} = (a_1, \ldots, a_N) \in \prod_{n=1}^{N} \mathcal{M}_n$ be the channel selection profile of all users. Given the channel selection profile $\boldsymbol{a}$, the interference received by user n can be computed as

$$\gamma_n(\boldsymbol{a}) = \sum_{m \in \mathcal{N}_n^p} P_m d_{mn}^{-\alpha} I_{\{a_n = a_m\}} + \omega_{a_n}^n. \quad (4.1)$$

Here α is the path loss factor and $I_{\{A\}}$ is an indicator function with $I_{\{A\}} = 1$ if the event A is true and $I_{\{A\}} = 0$ otherwise. Furthermore, $\omega^n_{a_n}$ denotes the noisy power including the interference from primary TV users on the channel a_n. We then define the individual utility function $u_n(\boldsymbol{a})$ as

$$u_n(\boldsymbol{a}) = -\gamma_n(\boldsymbol{a}) = -\sum_{m \in \mathcal{N}_n^p} P_m d_{mn}^{-\alpha} I_{\{a_n = a_m\}} - \omega_{a_n}^n. \quad (4.2)$$

Here the negative sign comes from the convention that utility functions are typically the ones to be maximized. The individual utility of user n reflects the fact that each user n has interest to reduce its own received interference. To capture the social coupling in the social graph $\mathcal{G}^s$, we further introduce the social group utility of each white-space user n according to (2.1) as

$$f_n(\boldsymbol{a}) = u_n(\boldsymbol{a}) + \sum_{m \in \mathcal{N}_n^s} s_{nm} u_m(\boldsymbol{a}). \quad (4.3)$$

We then formulate the database assisted spectrum access problem as a SGUM game $\Gamma = (\mathcal{N}, \{\mathcal{M}_n\}_{n\in\mathcal{N}}, \{f_n\}_{n\in\mathcal{N}})$, where the set of white-space users $\mathcal{N}$ is the set of players, the set of vacant channels $\mathcal{M}_n$ is the set of strategies for each player n, and the social group utility function f_n of each user n is the payoff function of player n.

4.3 Existence of Social-Aware Nash Equilibrium

We next study the existence of SNE of the SGUM game for database assisted spectrum access. Here we resort to a useful tool of potential game [5].

Definition 4.1 A game is called a potential game if it admits a potential function $\Phi(\boldsymbol{a})$ such that for every $n \in \mathcal{N}$ and $\boldsymbol{a}_{-n} \in \prod_{i\neq n} \mathcal{M}_i$, for any $a_n, a_n^{'} \in \mathcal{M}_n$,

$$f_n(a_n^{'}, \boldsymbol{a}_{-n}) - f_n(a_n, \boldsymbol{a}_{-n}) = \Phi(a_n^{'}, \boldsymbol{a}_{-n}) - \Phi(a_n, \boldsymbol{a}_{-n}). \tag{4.4}$$

An appealing property of the potential game is that it always admits a Nash equilibrium, and any strategy profile that maximizes the potential function $\Phi(\boldsymbol{a})$ is a Nash equilibrium [5].

For the SGUM game Γ for database assisted spectrum access, we can show that it is a potential game. For ease of exposition, we first introduce the *physical-social graph* $\mathcal{G}^{sp} = \{\mathcal{N}, \mathcal{E}^{sp}\}$ to capture both physical coupling and social coupling simultaneously. Here the vertex set is the same as the user set $\mathcal{N}$ and the edge set is given as $\mathcal{E}^{sp} = \{(n,m) : e_{nm}^{sp} \equiv e_{nm}^{s} \cdot e_{nm}^{p} = 1, \forall n, m \in \mathcal{N}\}$ where $e_{nm}^{sp} = 1$ if and only if users n and m have social tie between each other (i.e., $e_{nm}^{s} = 1$) and can also generate interference to each other (i.e., $e_{nm}^{p} = 1$). We denote the set of users that have social ties and can also generate interference to user n as $\mathcal{N}_n^{sp} = \{m : e_{nm}^{sp} = 1, \forall m \in \mathcal{N}\}$.

Based on the physical-social graph $\mathcal{G}^{sp}$, we show in Theorem 4.1 that the SGUM game Γ is a potential game with the following potential function

$$\Phi(\boldsymbol{a}) = \underbrace{-\frac{1}{2}\sum_{n=1}^{N}\sum_{m\in\mathcal{N}_n^p} P_m d_{mn}^{-\alpha} I_{\{a_n = a_m\}} - \sum_{n=1}^{N} \omega_{a_n}^{n}}_{\Phi_1(\boldsymbol{a})\text{: due to physical coupling}}$$

$$\underbrace{-\frac{1}{2}\sum_{n=1}^{N}\sum_{m\in\mathcal{N}_n^{sp}} s_{nm} P_m d_{mn}^{-\alpha} I_{\{a_n = a_m\}}}_{\Phi_2(\boldsymbol{a})\text{: due to social coupling}} . \tag{4.5}$$

The potential function in (4.5) can be decomposed into two parts: $\Phi_1(\boldsymbol{a})$ and $\Phi_2(\boldsymbol{a})$. The first part $\Phi_1(\boldsymbol{a})$ reflects the weighted system-wide interference level (including background noise) due to physical coupling in the physical domain and the second part $\Phi_2(\boldsymbol{a})$ captures the interdependence of user utilities due to social coupling in the social domain.

Theorem 4.1 *The SGUM game Γ for database assisted spectrum access is a potential game with the potential function $\boldsymbol{\Phi}(\boldsymbol{a})$ given in* (4.5)*, and hence has a SNE.*

The proof is given in Appendix. Note that when $s_{nm} = 0$ for any user $n, m \in \mathcal{N}$ (i.e., all users are selfish), the potential function $\Phi(\boldsymbol{a}) = \Phi_1(\boldsymbol{a})$, which does not involve the social coupling part $\Phi_2(\boldsymbol{a})$. In this case, the SGUM game Γ for database assisted spectrum access degenerates to the non-cooperative spectrum access game. When $s_{nm} = 1$ for any user $n, m \in \mathcal{N}$ (i.e., all users are fully altruistic), the potential function $\Phi(\boldsymbol{a}) = \sum_{n=1}^{N} u_n(\boldsymbol{a})$, which is the system-wide utility. In this case, the SGUM becomes the network utility maximization.

We next design a distributed spectrum access algorithm that can achieve the SNE of the SGUM game Γ for database assisted spectrum access.

4.4 Distributed Spectrum Access Algorithm

In this section we study the distributed spectrum access algorithm design.

4.4.1 Algorithm Design Principles

According to the property of potential game, any channel selection profile $\boldsymbol{a}$ that maximizes the potential function $\Phi(\boldsymbol{a})$ is a Nash equilibrium [5]. We hence design a distributed spectrum access algorithm that achieves the SNE of the SGUM Γ by maximizing the potential function $\Phi(\boldsymbol{a})$.

To proceed, first consider the problem that the users collectively compute the optimal channel selection profile such that the potential function is maximized, i.e.,

$$\max_{\boldsymbol{a} \in \Omega \equiv \prod_{n=1}^{N} \mathcal{M}_n} \Phi(\boldsymbol{a}). \tag{4.6}$$

The problem (4.6) involves a combinatorial optimization over the discrete solution space Ω. In general, it is very challenging to solve such a problem exactly especially when the system size is large (i.e., the solution space Ω is large).

With this observation, it is plausible to search for approximate solutions to the potential function maximization problem. To this end, rewrite

We then consider to approach the potential function maximization solution approximately. To proceed, we first write the problem (4.6) as the following equivalent randomized problem:

$$\max_{(q_{\boldsymbol{a}} \geq 0 : \boldsymbol{a} \in \Omega)} \sum_{\boldsymbol{a} \in \Omega} q_{\boldsymbol{a}} \Phi(\boldsymbol{a})$$

$$\text{s.t.} \sum_{\boldsymbol{a} \in \Omega} q_{\boldsymbol{a}} = 1, \tag{4.7}$$

where q_a is the probability that channel selection profile $\boldsymbol{a}$ is adopted. Obviously, the optimal solution to problem (4.7) is to choose the optimal channel selection profiles with probability one. It is known from the Markov approximation approach in [6] that problem (4.7) can be approximated by the following convex optimization problem:

$$\begin{aligned} \max_{(q_a \geq 0 : \boldsymbol{a} \in \Omega)} & \sum_{\boldsymbol{a} \in \Omega} q_a \Phi(\boldsymbol{a}) - \frac{1}{\theta} \sum_{\boldsymbol{a} \in \Omega} q_a \ln q_a \\ \text{s.t.} & \sum_{\boldsymbol{a} \in \Omega} q_a = 1, \end{aligned} \tag{4.8}$$

where θ is the parameter that controls the approximation ratio. Note that the approximation in (4.8) can guarantee the asymptotic optimality. This is because that when $\theta \to \infty$, the problem (4.8) boils down to exactly the same as problem (4.7). That is, when $\theta \to \infty$, the optimal solutions that maximize the potential function $\Phi(\boldsymbol{a})$ will be selected with probability one. Moreover, the approximation in (4.8) enables us to obtain the close-form solution, which facilitates the distributed algorithm design later. More specifically, by the KKT conditions [7], the optimal solution to problem (4.8) is given as

$$q_a^* = \frac{\exp(\theta \Phi(\boldsymbol{a}))}{\sum_{\hat{\boldsymbol{a}} \in \Omega} \exp(\theta \Phi(\hat{\boldsymbol{a}}))}. \tag{4.9}$$

Based on (4.9), we then design a self-organizing algorithm such that the asynchronous channel selection updates of the users form a Markov chain (with the system state as the channel selection profile $\boldsymbol{a}$ of all users). As long as the Markov chain converges to the stationary distribution as given in (4.9), we can approach the Nash equilibrium channel selection profile that maximizes the potential function by setting a large enough parameter θ.

4.4.2 *Markov Chain Design for Distributed Spectrum Access*

Motivated by the seminal work on the adaptive CSMA mechanism [8], we propose a distributed spectrum access algorithm in Algorithm 1 such that each user n updates its channel selection according to a timer value that follows the exponential distribution with a rate of τ_n. Note that the study in [8] focuses on the network utility maximization, while in this paper we consider the social group utility maximization, which results in significant differences in analysis.

Appealing to the property of exponential distributions, we have that the probability that more than one users generate the same timer value and update their channels simultaneously equals zero. Since one user will activate for the channel selection update at a time, the direct transitions between two system states $\boldsymbol{a}$ and $\boldsymbol{a}'$ are feasible if these two system states differ by one and only one user's channel selection. We also denote the set of system states that can be transited directly from the state $\boldsymbol{a}$

Algorithm 1 Distributed Spectrum Access Algorithm For Social Group Utility Maximization

initialization:
  set the parameter θ and the channel update rate τ_n.
  choose a channel $a_n \in \mathcal{M}_n$ randomly for each user $n \in \mathcal{N}$.
end initialization

loop for each user $n \in \mathcal{N}$ in parallel:
  compute the social group utility $f_n(a_n, \boldsymbol{a}_{-n})$ on the chosen channel a_n.
  generate a timer value following the exponential distribution with the mean equal to $\frac{1}{\tau_n}$.
  count down until the timer expires.
  if the timer expires **then**
    choose a new channel $a_n^{'} \in \mathcal{M}_n$ randomly.
    compute the social group utility $f_n(a_n^{'}, \boldsymbol{a}_{-n})$ on the new channel $a_n^{'}$.
    stay in the new channel $a_n^{'}$ with probability $\frac{\exp\left(\theta f_n(a_n^{'}, \boldsymbol{a}_{-n})\right)}{\max\{\exp\left(\theta f_n(a_n^{'}, \boldsymbol{a}_{-n})\right), \exp(\theta f_n(a_n, \boldsymbol{a}_{-n}))\}}$, Or
  move back to the original channel a_n with probability $1 - \frac{\exp\left(\theta f_n(a_n^{'}, \boldsymbol{a}_{-n})\right)}{\max\{\exp\left(\theta f_n(a_n^{'}, \boldsymbol{a}_{-n})\right), \exp(\theta f_n(a_n, \boldsymbol{a}_{-n}))\}}$.
  end if
end loop

as $\Delta_{\boldsymbol{a}} = \{\boldsymbol{a}^{'} \in \Omega : |\{\boldsymbol{a}^{'} \cup \boldsymbol{a}\} \backslash \{\boldsymbol{a}^{'} \cap \boldsymbol{a}\}| = 2\}$, where $|\cdot|$ denotes the size of a set. According to (4.3), a user n can compute the social group utility $f_n(\boldsymbol{a})$ by locally enquiring the users having social ties with it about their received interferences. Then user n will randomly choose a new channel $a_n^{'} \in \mathcal{M}_n$ and stay in this channel with a probability of

$$\frac{\exp\left(\theta f_n(a_n^{'}, \boldsymbol{a}_{-n})\right)}{\max\{\exp\left(\theta f_n(a_n^{'}, \boldsymbol{a}_{-n})\right), \exp\left(\theta f_n(a_n, \boldsymbol{a}_{-n})\right)\}}. \tag{4.10}$$

The underlying intuition behind (4.10) is as follows. When $f_n(a_n^{'}, \boldsymbol{a}_{-n}) \geq f_n(a_n, \boldsymbol{a}_{-n})$ (i.e., the new channel $a_n^{'}$ offers the better performance), user n will stay in the new channel $a_n^{'}$ with probability one. According to the property of potential game in (4.4), we know that choosing the new channel $a_n^{'}$ can help to increase both user n's social group utility $f_n(\boldsymbol{a})$ and the potential function $\Phi(\boldsymbol{a})$ of the SGUM game. When $f_n(a_n^{'}, \boldsymbol{a}_{-n}) < f_n(a_n, \boldsymbol{a}_{-n})$ (i.e., the original channel a_n offers the better performance), user n will switch back to the original channel a_n with a probability of $1 - \frac{\exp\left(\theta f_n(a_n^{'}, \boldsymbol{a}_{-n})\right)}{\exp(\theta f_n(a_n, \boldsymbol{a}_{-n}))}$. That is, the probability that user n will switch back to the original channel a_n increases with the performance gap $f_n(a_n, \boldsymbol{a}_{-n}) - f_n(a_n^{'}, \boldsymbol{a}_{-n})$. We would like to emphasize that such probabilistic channel selections are necessary such that all the users can explore the feasible channel selection space to prevent the algorithm from getting stuck at a local optimum.

Then from a system-wide perspective, the probability of transition from state $(a_n, \boldsymbol{a}_{-n})$ to $(a_n^{'}, \boldsymbol{a}_{-n})$ due to user n's channel selection update is given as

$$\frac{1}{|\mathcal{M}_n|} \frac{\exp\left(\theta f_n(a_n^{'}, \boldsymbol{a}_{-n})\right)}{\max\{\exp\left(\theta f_n(a_n^{'}, \boldsymbol{a}_{-n})\right), \exp\left(\theta f_n(a_n, \boldsymbol{a}_{-n})\right)\}}. \tag{4.11}$$

Since each user n activates its channel selection update according to the countdown timer mechanism with a rate of τ_n, hence if $\boldsymbol{a}^{'} \in \Delta_{\boldsymbol{a}}$, the transition rate from state $\boldsymbol{a}$ to state $\boldsymbol{a}^{'}$ is given as

$$q_{\boldsymbol{a},\boldsymbol{a}'} = \frac{\tau_n}{|\mathcal{M}_n|} \frac{\exp\left(\theta f_n(a_n^{'}, \boldsymbol{a}_{-n})\right)}{\max\{\exp\left(\theta f_n(a_n^{'}, \boldsymbol{a}_{-n})\right), \exp\left(\theta f_n(a_n, \boldsymbol{a}_{-n})\right)\}}. \tag{4.12}$$

Otherwise, we have $q_{\boldsymbol{a},\boldsymbol{a}'} = 0$. We show in Theorem 4.2 that the spectrum access Markov chain is time reversible. Time reversibility means that when tracing the Markov chain backwards, the stochastic behavior of the reverse Markov chain remains the same. A nice property of a time reversible Markov chain is that it always admits a unique stationary distribution, which is independent of the initial system state. This implies that given any initial channel selections the distributed spectrum access algorithm can drive the system converging to the stationary distribution given in (4.9).

Theorem 4.2 *The distributed spectrum access algorithm induces a time-reversible Markov chain with the unique stationary distribution as given in* (4.9).

The proof is given in Appendix. One key idea of the proof is to show that the distribution in (4.9) satisfies the following detailed balance equations: $q_{\boldsymbol{a}}^{*} q_{\boldsymbol{a},\boldsymbol{a}'} = q_{\boldsymbol{a}'}^{*} q_{\boldsymbol{a}',\boldsymbol{a}}, \forall \boldsymbol{a}, \boldsymbol{a}^{'} \in \Omega$.

4.4.3 Performance Analysis

According to Theorem 4.2, we can achieve the SNE that maximizes the potential function $\Phi(\boldsymbol{a})$ of the SGUM game Γ by setting $\theta \rightarrow \infty$. However, in practice one has to choose only implement a finite value of θ. Let $\bar{\Phi} = \sum_{\boldsymbol{a}\in\Omega} q_{\boldsymbol{a}}^{*}\Phi(\boldsymbol{a})$ be the expected potential by Algorithm 1 and $\Phi^{*} = \max_{\boldsymbol{a}\in\Omega}\Phi(\boldsymbol{a})$ be the maximum potential. We show in Theorem 4.3 that, when a large enough θ is adopted in practice, the gap between $\bar{\Phi}$ and Φ^{*} is very small.

Theorem 4.3 *For the distributed spectrum access algorithm, we have that* $0 \leq \Phi^{*} - \bar{\Phi} \leq \frac{1}{\theta}\sum_{n=1}^{N} \ln|\mathcal{M}_n|$, *where* $|\mathcal{M}_n|$ *denotes the number of vacant channels of user* n.

Proof First of all, we must have that $\Phi^{*} \geq \bar{\Phi}$. According to (4.7) and (4.8), we then have that

$$\max_{(q_{\boldsymbol{a}}:\boldsymbol{a}\in\Omega)} \sum_{\boldsymbol{a}\in\Omega} q_{\boldsymbol{a}}\Phi(\boldsymbol{a}) \leq \max_{(q_{\boldsymbol{a}}:\boldsymbol{a}\in\Omega)} \sum_{\boldsymbol{a}\in\Omega} q_{\boldsymbol{a}}\Phi(\boldsymbol{a}) - \frac{1}{\theta}\sum_{\boldsymbol{a}\in\Omega} q_{\boldsymbol{a}} \ln q_{\boldsymbol{a}}, \tag{4.13}$$

which is due to the fact that $0 \leq -\frac{1}{\theta}\sum_{\boldsymbol{a}\in\Omega} q_{\boldsymbol{a}} \ln q_{\boldsymbol{a}} \leq \frac{1}{\theta}\ln|\Omega|$. Since $q_{\boldsymbol{a}}^{*}$ is the optimal solution to (4.8) and $\Phi^{*} = \max_{(q_{\boldsymbol{a}}:\boldsymbol{a}\in\Omega)} \sum_{\boldsymbol{a}\in\Omega} q_{\boldsymbol{a}}\Phi(\boldsymbol{a})$, according to (4.13), we know that

$$\Phi^{*} \leq \sum_{\boldsymbol{a}\in\Omega} q_{\boldsymbol{a}}^{*}\Phi(\boldsymbol{a}) - \frac{1}{\theta}\sum_{\boldsymbol{a}\in\Omega} q_{\boldsymbol{a}}^{*} \ln q_{\boldsymbol{a}}^{*}$$

$$\leq \sum_{a\in\Omega} q_a^* \Phi(\boldsymbol{a}) + \frac{1}{\theta}\ln|\Omega| \leq \bar{\Phi} + \frac{1}{\theta}\ln|\Omega|,$$

which completes the proof. □

We next discuss the efficiency of the SNE by the distributed spectrum access algorithm when θ is sufficiently large (i.e., $\theta \rightarrow \infty$). Let $V(\boldsymbol{a})$ be the total individual utility received by all the users under the channel selection profile $\boldsymbol{a}$, i.e., $V(\boldsymbol{a}) = \sum_{n=1}^{N} u_n(\boldsymbol{a})$. We denote $\overline{\boldsymbol{a}}$ as the NUM solution that maximizes the system-wide utility (i.e., $\overline{\boldsymbol{a}} = \arg\max_{a\in\Omega} V(\boldsymbol{a})$) and $\hat{\boldsymbol{a}}$ as the convergent SNE by the distributed spectrum access algorithm (i.e., $\hat{\boldsymbol{a}} = \arg\max_{a\in\Omega} \Phi(\boldsymbol{a})$). We then define the performance gap ρ as the difference between the total utility received at the NUM solution $\overline{\boldsymbol{a}}$ and that of the SNE $\hat{\boldsymbol{a}}$, i.e., $\rho = V(\overline{\boldsymbol{a}}) - V(\hat{\boldsymbol{a}})$. We can show the following result.

Theorem 4.4 *The performance gap ρ of the SNE by the distributed spectrum access algorithm is at most*

$$\frac{1}{2}\sum_{n=1}^{N}\sum_{m\in\mathcal{N}_n^{sp}} (1-s_{nm})P_m d_{mn}^{-\alpha} + \frac{1}{2}\sum_{n=1}^{N}\sum_{m\in\mathcal{N}_n^{p}\setminus\mathcal{N}_n^{sp}} P_m d_{mn}^{-\alpha}.$$

Proof According to (4.1) and (4.5), we have that

$$V(\boldsymbol{a}) = \sum_{n=1}^{N} u_n(\boldsymbol{a}) = -\sum_{n=1}^{N}\sum_{m\in\mathcal{N}_n^{p}} P_m d_{mn}^{-\alpha} I_{\{a_n=a_m\}} - \sum_{n=1}^{N}\omega_{a_n}^{n}$$
$$= \Phi(\boldsymbol{a}) - \frac{1}{2}\sum_{n=1}^{N}\sum_{m\in\mathcal{N}_n^{sp}} (1-s_{nm})P_m d_{mn}^{-\alpha} I_{\{a_n=a_m\}}$$
$$- \frac{1}{2}\sum_{n=1}^{N}\sum_{m\in\mathcal{N}_n^{p}\setminus\mathcal{N}_n^{sp}} P_m d_{mn}^{-\alpha} I_{\{a_n=a_m\}}.$$

We then have that

$$\rho = V(\overline{\boldsymbol{a}}) - V(\hat{\boldsymbol{a}}) = \Phi(\overline{\boldsymbol{a}}) - \Phi(\hat{\boldsymbol{a}})$$
$$- \frac{1}{2}\sum_{n=1}^{N}\sum_{m\in\mathcal{N}_n^{sp}} (1-s_{nm})P_m d_{mn}^{-\alpha}\left(I_{\{\overline{a}_n=\overline{a}_m\}} - I_{\{\hat{a}_n=\hat{a}_m\}}\right)$$
$$- \frac{1}{2}\sum_{n=1}^{N}\sum_{m\in\mathcal{N}_n^{p}\setminus\mathcal{N}_n^{sp}} P_m d_{mn}^{-\alpha}\left(I_{\{\overline{a}_n=\overline{a}_m\}} - I_{\{\hat{a}_n=\hat{a}_m\}}\right). \quad (4.14)$$

Since $\Phi(\hat{\boldsymbol{a}}) = \max_{a\in\Omega}\Phi(\boldsymbol{a}) \geq \Phi(\overline{\boldsymbol{a}})$ and $V(\overline{\boldsymbol{a}}) = \max_{a\in\Omega} V(\boldsymbol{a}) \geq V(\hat{\boldsymbol{a}})$, we know from (4.14) that

$$\rho \leq \frac{1}{2}\sum_{n=1}^{N}\sum_{m\in\mathcal{N}_n^{sp}} (1-s_{nm})P_m d_{mn}^{-\alpha}\left(I_{\{\hat{a}_n=\hat{a}_m\}} - I_{\{\overline{a}_n=\overline{a}_m\}}\right)$$

$$+\frac{1}{2}\sum_{n=1}^{N}\sum_{m\in\mathcal{N}_n^p\setminus\mathcal{N}_n^{sp}} P_m d_{mn}^{-\alpha}\left(I_{\{\hat{a}_n=\hat{a}_m\}}-I_{\{\bar{a}_n=\bar{a}_m\}}\right)$$

$$\leq\frac{1}{2}\sum_{n=1}^{N}\sum_{m\in\mathcal{N}_n^{sp}}(1-s_{nm})P_m d_{mn}^{-\alpha}+\frac{1}{2}\sum_{n=1}^{N}\sum_{m\in\mathcal{N}_n^p\setminus\mathcal{N}_n^{sp}} P_m d_{mn}^{-\alpha}. \qquad \square$$

Theorem 4.4 indicates that the upper-bound of the performance gap ρ decreases as the strength of social tie s_{nm} among users increases. When $s_{nm}=0$ for any user $n, m \in \mathcal{N}$ (i.e., all users are selfish), the social group utility maximization game Γ degenerates to the non-cooperative spectrum access game and the upper-bound of the performance gap ρ reaches the maximum of $\frac{1}{2}\sum_{n=1}^{N}\sum_{m\in\mathcal{N}_n^p} P_m d_{mn}^{-\alpha}$. When $s_{nm}=1$ for any user $n, m \in \mathcal{N}$ (i.e., all users are fully altruistic), the SGUM becomes the NUM and the performance gap $\rho=0$. In Sect. 4.5, we also evaluate the performance of the SGUM solution by real social data traces. Numerical results demonstrate that the performance gap between the SGUM solution and the NUM solution is at most 15 %.

4.5 Numerical Results

In this section, we evaluate the SGUM solution for database assisted spectrum access by numerical studies based on both Erdos-Renyi social graphs and real trace based social graphs.

4.5.1 Social Graph with 8 White-Space Users

We first consider a database assisted spectrum access network consisting of $M=5$ channels and $N=8$ white-space users, which are scattered across a square area of a length of 500 m (see Fig. 4.2). The transmission power of each user is $P_n=100$ mW [1], the path loss factor $\alpha=4$, and the background interference power ω_m^n for each channel m and user n is randomly assigned in the interval of $[-100,-90]$ dBm. Each user n has a different set of vacant channels by consulting the geo-location database. For example, the vacant channels for user 1 are $\{2,3,4\}$. For the interference graph $\mathcal{G}^p$, we define that the user's transmission range $\delta=1000$ m and two users can generate inference to each other if their distance is not greater than δ. The social graph $\mathcal{G}^s$ is given in Fig. 4.2 where two users have social tie if there is an edge between them and the numerical value associated with each edge represents the strength of social tie.

We implement the proposed distributed spectrum access algorithm for the SGUM game with different parameters θ in Fig. 4.3. We see that the convergent potential function value Φ of the SGUM game increases as the parameter θ increases. When the parameter θ is large enough (e.g., $\theta \geq 10^6$), the algorithm can approach the maximum potential function value $\Phi^*=\max_{\boldsymbol{a}}\Phi(\boldsymbol{a})$. Figure 4.4 shows the dynamics of user's

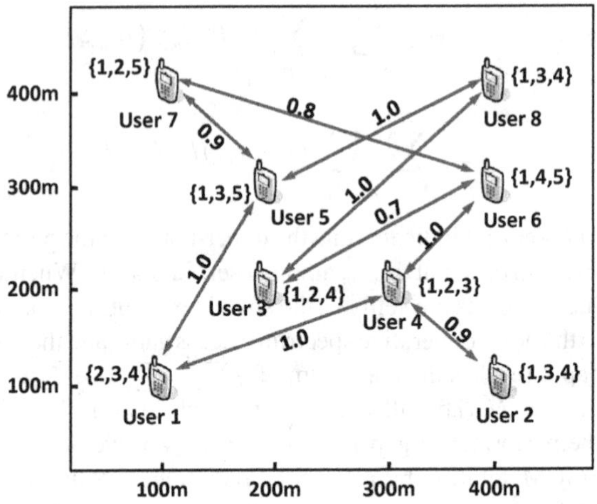

Fig. 4.2 A square area of a length of 500 m with 8 scattered white-space users

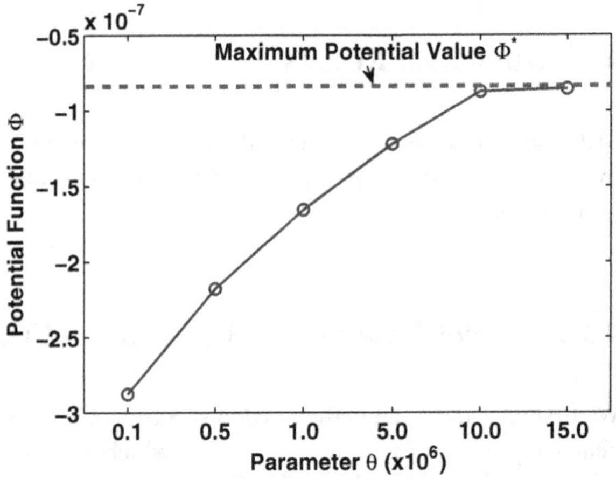

Fig. 4.3 The convergent potential value Φ with different parameters θ

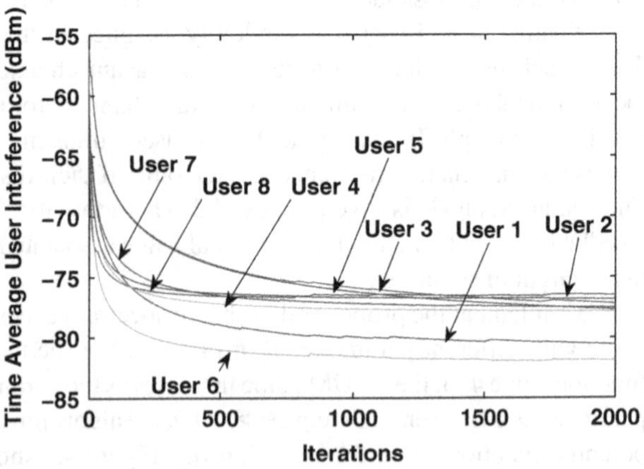

Fig. 4.4 Dynamics of users' time average interference when $\theta = 10^6$

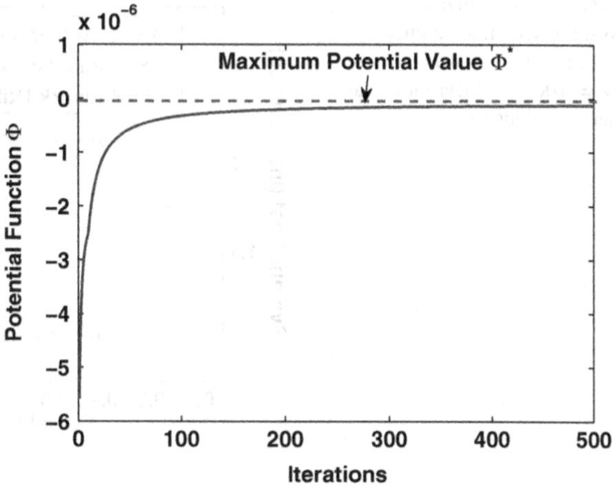

Fig. 4.5 Dynamics of potential value Φ when $\theta = 10^6$

time average interference $\gamma_n(\boldsymbol{a})$. It demonstrates that the distributed spectrum access algorithm can drive users' time average interference decreasing and converging to an equilibrium such that each user only receives a small interference level. To verify that the algorithm can approach the SNE of the SGUM game, we show the dynamics of the potential value $\Phi(\boldsymbol{a})$ in Fig. 4.5. We see that the distributed spectrum access algorithm can drive the potential value Φ increasing and approach the maximum potential value Φ^*. According to the property of potential game, the algorithm hence can approach the SNE of the SGUM game.

4.5.2 Erdos-Renyi Social Graph

We then consider $N = 100$ users that randomly scattered across a square area of a length of 2000 m. We evaluate the SGUM game solution by the distributed spectrum access algorithm with the social graph represented by the Erdos-Renyi (ER) graph model [9], where a social link exists between any two users with a probability of P_L. We set the strength of social tie $s_{nm} = 1$ for each social link. To evaluate the impact of social link density of the social graph, we implement the simulations with different social link probabilities $P_L = 0, 0.1, \ldots, 1.0$, respectively. For each given P_L, we average over 100 runs. To benchmark the SGUM solution, we also implement the the following two solutions:

(1) Non-cooperative spectrum access: we implement the non-cooperative game based solution such that each user aims to maximize its individual utility, i.e., we set $f_n(\boldsymbol{a}) = u_n(\boldsymbol{a})$ in the distributed spectrum access algorithm.
(2) Network utility maximization: we implement the social optimal solution such that the system-wide utility is maximized, i.e., we set $f_n(\boldsymbol{a}) = \sum_{n=1}^{N} u_n(\boldsymbol{a})$ in the distributed spectrum access algorithm.

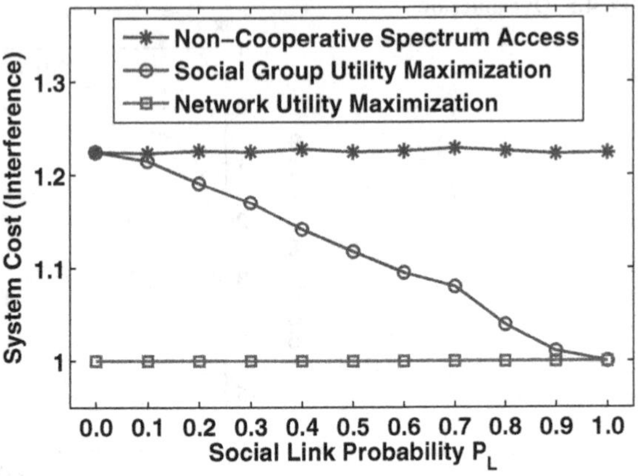

Fig. 4.6 Normalized system-wide interference with number of nodes $N = 100$ and different social network density

Similar to the price of anarchy in non-cooperative game [10], we normalize the system-wide interference of these solutions with respect to that of the social optimal solution (i.e., network utility maximum solution). The results are given in Fig. 4.6. We see that the performance of the SGUM solution always dominates that of the non-cooperative spectrum access. This is non-trivial since non-cooperative game promotes the competition among users to increase the system-wide utility and has been widely adopted to devise efficient distributed resource allocation mechanisms in wireless networks [11]. Moreover, we observe that the performance gain of the SGUM solution increases as the social link probability P_L increases. When the social link probability $P_L = 1$, the SGUM solution achieves the same performance of the network utility maximization and can reduce 23 % system-wide interference over the non-cooperative spectrum access. This also demonstrates that the proposed SGUM framework spans the continuum space between non-cooperative game and network utility maximization—two extreme paradigms based on drastically different assumptions that users are selfish and altruistic, respectively.

4.5.3 Real Trace Based Social Graph

We next evaluate the SGUM solution by the distributed spectrum access algorithm based on the social graph represented by the friendship network of the real data trace Brightkite [12]. We implement experiments with the number of users $N = 200, 300, \ldots, 600$, respectively. As the benchmark, we also implement the solutions of non-cooperative spectrum access and network utility maximization.

The results are shown in Fig. 4.7. We see that the non-cooperative spectrum access solution will increase the system-wide interference up-to 29 % over the network utility maximization solution. Upon comparison, the system-wide interference by the the SGUM solution will increase at most 15 %, compared with the network utility maximization solution. This verifies the effectiveness of leveraging social tie to stimulate user cooperation for achieving efficient distributed spectrum access in practices.

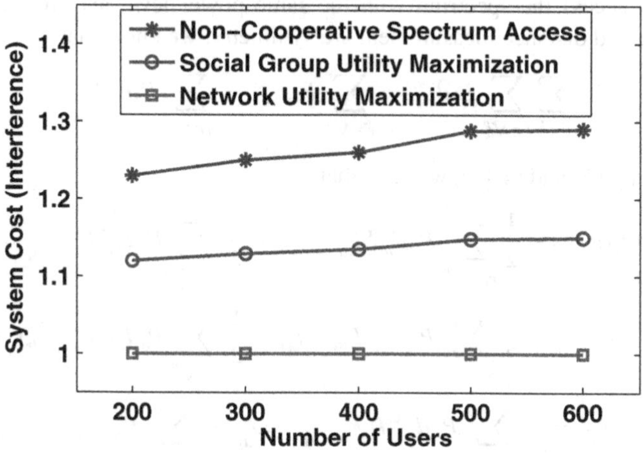

Fig. 4.7 Normalized system-wide interference with different number of users

4.6 Summary

In this chapter, we study the SGUM-based database assisted spectrum access. We show that the SGUM-based spectrum access game is a potential game and thus always admits a SNE. Then we design a distributed spectrum access algorithm that can achieve an SNE. We also derive the upper-bound of the performance gap of the SNE from the NUM solution. Numerical results demonstrate that the performance gap between the SGUM solution and the NUM solution is at most 15 %.

Appendix

Proof of Theorem 4.1

Suppose that a user k changes its channel a_k to $a_k^{'}$ such that the channel selection profile changes from $\boldsymbol{a}$ to $\boldsymbol{a}^{'}$. We have that

$$\Phi(\boldsymbol{a}^{'}) - \Phi(\boldsymbol{a}) = \Phi_1(\boldsymbol{a}^{'}) - \Phi_1(\boldsymbol{a}) + \Phi_2(\boldsymbol{a}^{'}) - \Phi_2(\boldsymbol{a}). \tag{4.15}$$

For the part Φ_1, we have that

$$\begin{aligned}\Phi_1(\boldsymbol{a}^{'}) - \Phi_1(\boldsymbol{a}) = &-\frac{1}{2}\sum_{m\in\mathcal{N}_k^P} P_m d_{mk}^{-\alpha} I_{\{a_k^{'}=a_m\}} - \frac{1}{2}\sum_{n\neq k}\sum_{k\in\mathcal{N}_n^P} P_k d_{kn}^{-\alpha} I_{\{a_n=a_k^{'}\}} \\ &- \omega_{a_k^{'}}^{k} + \frac{1}{2}\sum_{m\in\mathcal{N}_k^P} P_m d_{mk}^{-\alpha} I_{\{a_k=a_m\}} \\ &+ \frac{1}{2}\sum_{n\neq k}\sum_{k\in\mathcal{N}_n^P} P_k d_{kn}^{-\alpha} I_{\{a_n=a_k\}} + \omega_{a_k}^{k}.\end{aligned} \tag{4.16}$$

Since users access the spectrum with the same power level and the interference relationship and distance measurement are symmetry, we know that

$$\sum_{n\neq k}\sum_{k\in\mathcal{N}_n^p} P_k d_{kn}^{-\alpha} = \sum_{n\in\mathcal{N}_k^p} P_k d_{kn}^{-\alpha} = \sum_{n\in\mathcal{N}_k^p} P_n d_{nk}^{-\alpha}. \tag{4.17}$$

Combining (4.16) and (4.17), we have that

$$\begin{aligned}
\Phi_1(\boldsymbol{a}^{'}) - \Phi_1(\boldsymbol{a}) =& -\frac{1}{2}\sum_{m\in\mathcal{N}_k^p} P_m d_{mk}^{-\alpha} I_{\{a_k^{'} = a_m\}} - \frac{1}{2}\sum_{n\in\mathcal{N}_k^p} P_n d_{nk}^{-\alpha} I_{\{a_n = a_k^{'}\}} - \omega_{a_k^{'}}^k \\
&+\frac{1}{2}\sum_{m\in\mathcal{N}_k^p} P_m d_{mk}^{-\alpha} I_{\{a_k = a_m\}} + \frac{1}{2}\sum_{n\in\mathcal{N}_k^p} P_n d_{nk}^{-\alpha} I_{\{a_n = a_k\}} + \omega_{a_k}^k \\
=& -\sum_{m\in\mathcal{N}_k^p} P_m d_{mk}^{-\alpha} I_{\{a_k^{'} = a_m\}} - \omega_{a_k^{'}}^k + \sum_{m\in\mathcal{N}_k^p} P_m d_{mk}^{-\alpha} I_{\{a_k = a_m\}} + \omega_{a_k}^k \\
=& -\gamma_k(\boldsymbol{a}^{'}) + \gamma_k(\boldsymbol{a}) = U_k(\boldsymbol{a}^{'}) - U_k(\boldsymbol{a}).
\end{aligned} \tag{4.18}$$

Similarly, for the part Φ_2, we have that

$$\begin{aligned}
\Phi_2(\boldsymbol{a}^{'}) - \Phi_2(\boldsymbol{a}) =& \sum_{n\in\mathcal{N}_k^{sp}} w_{kn}\left(-P_k d_{kn}^{-\alpha} I_{\{a_n = a_k^{'}\}} + P_k d_{kn}^{-\alpha} I_{\{a_n = a_k\}}\right) \\
=& \sum_{n\in\mathcal{N}_k^{sp}} w_{kn} \times \\
&\left(-P_k d_{kn}^{-\alpha} I_{\{a_n = a_k^{'}\}} - \sum_{m\neq k}\sum_{m\in\mathcal{N}_n^p} P_m d_{mn}^{-\alpha} I_{\{a_n = a_m\}} - \omega_{a_n}^k \right.\\
&\left. + P_k d_{kn}^{-\alpha} I_{\{a_n = a_k\}} + \sum_{m\neq k}\sum_{m\in\mathcal{N}_n^p} P_m d_{mn}^{-\alpha} I_{\{a_n = a_m\}} + \omega_{a_n}^k\right) \\
=& \sum_{n\in\mathcal{N}_k^{sp}} w_{kn}\left(-\gamma_n(\boldsymbol{a}^{'}) + \gamma_n(\boldsymbol{a})\right) \\
=& \sum_{n\in\mathcal{N}_k^{sp}} w_{kn}\left(U_n(\boldsymbol{a}^{'}) - U_n(\boldsymbol{a})\right).
\end{aligned} \tag{4.19}$$

Finally, substituting (4.18) and (4.19) into (4.15), we obtain that

$$\Phi(\boldsymbol{a}^{'}) - \Phi(\boldsymbol{a}) = U_k(\boldsymbol{a}^{'}) - U_k(\boldsymbol{a}) + \sum_{n\in\mathcal{N}_k^{sp}} w_{kn}\left(U_n(\boldsymbol{a}^{'}) - U_n(\boldsymbol{a})\right). \tag{4.20}$$

Since user k can not generate interference to any user $n \in \mathcal{N}_k^s \backslash \mathcal{N}_k^{sp}$, we have that

$$U_n(\boldsymbol{a}^{'}) = U_n(\boldsymbol{a}), \forall n \in \mathcal{N}_k^s \backslash \mathcal{N}_k^{sp}.$$

This implies that

$$\begin{aligned}\Phi(\boldsymbol{a}^{'}) - \Phi(\boldsymbol{a}) &= U_k(\boldsymbol{a}^{'}) - U_k(\boldsymbol{a}) + \sum_{n \in \mathcal{N}_k^{sp}} w_{kn}\left(U_n(\boldsymbol{a}^{'}) - U_n(\boldsymbol{a})\right) \\ &\quad + \sum_{n \in \mathcal{N}_k^{s} \setminus \mathcal{N}_k^{sp}} w_{kn}\left(U_n(\boldsymbol{a}^{'}) - U_n(\boldsymbol{a})\right) \\ &= U_k(\boldsymbol{a}^{'}) - U_k(\boldsymbol{a}) + \sum_{n \in \mathcal{N}_k^{s}} w_{kn}\left(U_n(\boldsymbol{a}^{'}) - U_n(\boldsymbol{a})\right),\end{aligned}$$

which completes the proof. □

Proof of Theorem 4.2

As mentioned, the system state of the spectrum access Markov chain is defined as the channel selection profile $\boldsymbol{a} \in \Theta$ of all users. Since it is possible to get from any state to any other state within finite steps of transition, the spectrum access Markov chain is hence irreducible and has a stationary distribution.

We then show that the Markov chain is time reversible by showing that the distribution in (4.9) satisfies the following detailed balance equations:

$$q_{\boldsymbol{a}}^{*} q_{\boldsymbol{a},\boldsymbol{a}^{'}} = q_{\boldsymbol{a}^{'}}^{*} q_{\boldsymbol{a}^{'},\boldsymbol{a}}, \forall \boldsymbol{a}, \boldsymbol{a}^{'} \in \Theta. \tag{4.21}$$

To see this, we consider the following two cases:

1) If $\boldsymbol{a}^{'} \notin \Delta_{\boldsymbol{a}}$, we have $q_{\boldsymbol{a},\boldsymbol{a}^{'}} = q_{\boldsymbol{a}^{'},\boldsymbol{a}} = 0$ and the Eq. (4.21) holds.

2) If $\boldsymbol{a}^{'} \in \Delta_{\boldsymbol{a}}$, according to (4.9) and (4.12), we have

$$\begin{aligned}q_{\boldsymbol{a}}^{*} q_{\boldsymbol{a},\boldsymbol{a}^{'}} &= \frac{\tau_n}{|\mathcal{M}_n|} \frac{\exp(\theta \Phi(\boldsymbol{a}))}{\sum_{\hat{\boldsymbol{a}} \in \Theta} \exp(\theta \Phi(\hat{\boldsymbol{a}}))} \\ &\quad \times \frac{\exp\left(\theta S_n(a_n^{'}, a_{-n})\right)}{\max\{\exp\left(\theta S_n(a_n^{'}, a_{-n})\right), \exp\left(\theta S_n(a_n, a_{-n})\right)\}} \\ &= \frac{\tau_n}{|\mathcal{M}_n|} \frac{\exp\left(\theta \left(\Phi(\boldsymbol{a}) + S_n(a_n^{'}, a_{-n})\right)\right)}{\sum_{\hat{\boldsymbol{a}} \in \Theta} \exp(\theta \Phi(\hat{\boldsymbol{a}}))} \\ &\quad \times \frac{1}{\max\{\exp\left(\theta S_n(a_n^{'}, a_{-n})\right), \exp\left(\theta S_n(a_n, a_{-n})\right)\}},\end{aligned}$$

and similarly,

$$\begin{aligned}q_{\boldsymbol{a}^{'}}^{*} q_{\boldsymbol{a}^{'},\boldsymbol{a}} &= \frac{\tau_n}{|\mathcal{M}_n|} \frac{\exp\left(\theta \left(\Phi(\boldsymbol{a}^{'}) + S_n(a_n, a_{-n})\right)\right)}{\sum_{\hat{\boldsymbol{a}} \in \Theta} \exp(\theta \Phi(\hat{\boldsymbol{a}}))} \\ &\quad \times \frac{1}{\max\{\exp\left(\theta S_n(a_n^{'}, a_{-n})\right), \exp\left(\theta S_n(a_n, a_{-n})\right)\}}.\end{aligned}$$

Thus, according to (4.4), we must have

$$q_a^* q_{a,a'} = q_{a'}^* q_{a',a}.$$

The spectrum access Markov chain is hence time-reversible and has the unique stationary distribution as given in (4.9). □

References

1. FCC, *Third Memorandum Opinion and Order*, FCC Std., April 5, 2012
2. L. Yang, H. Kim, J. Zhang, M. Chiang, C.W. Tan, Pricing-based decentralized spectrum access control in cognitive radio networks. IEEE/ACM Trans. Netw. **21**(2), 522–535, (2013)
3. X. Chen, J. Huang, Database-assisted distributed spectrum sharing. IEEE J. Sel. Areas Commun. **31**(11), 2349–2361, (2013)
4. K.N. Ramachandran, E.M. Belding-Royer, K.C. Almeroth, M.M. Buddhikot, Interference-aware channel assignment in multi-radio wireless mesh networks. INFOCOM **6**, 1–12 (2006)
5. D. Monderer, L.S. Shapley, Potential games. Games Econ. Behav. **14**(1), 124–143, (1996)
6. M. Chen, S.C. Liew, Z. Shao, C. Kai, Markov approximation for combinatorial network optimization. IEEE INFOCOM 1–9 (2010) (IEEE)
7. S. Boyd, L. Vandenberghe, *Convex Optimization* (Cambridge University Press, Cambridge, 2004)
8. L. Jiang, J. Walrand, A distributed CSMA algorithm for throughput and utility maximization in wireless networks. IEEE/ACM TON **18**(3), 960–972, (2010)
9. M.E. Newman, D.J. Watts, S.H. Strogatz, Random graph models of social networks. PANS **99**(Suppl 1), 2566–2572, (2002)
10. T. Roughgarden, *Selfish Routing and the Price of Anarchy* (MIT Press, Cambridge, 2005)
11. Z. Han, D. Niyato, W. Saad, T. Basar, A. Hjorungnes, *Game Theory in Wireless and Communication Networks* (Cambridge University Press, Cambridge, 2012)
12. E. Cho, S.A. Myers, J. Leskovec, Friendship and mobility: User movement in location-based social networks. ACM SIGKDD 1082–1090 (2011)

Chapter 5
SGUM-based Pseudonym Change for Personalized Location Privacy

In this chapter, we study the application of the SGUM framework to pseudonym change for personalized location privacy.

5.1 Introduction

With the rapid growth of mobile networks, location-based services (LBS) have become increasingly popular recently (e.g., location-based navigation and recommendation). However, the providers of LBSs are often considered not trustworthy, due to the risk of leaking users' location information to other parties (e.g., sell users' location data). As a result, mobile users are exposed to potential privacy threats when using a LBS. Although a user can use a pseudonym for the LBS, an adversary can infer the user's real identity from its location traces (e.g., from the user's home and work addresses). To protect location privacy, an effective approach is to "confuse" the adversary using the notion of *anonymity* [1]: mobile users in physical proximity can change their pseudonyms simultaneously to form an anonymity set, so that the adversary cannot distinguish any of them from the others.

A basic assumption commonly used in existing studies [2, 3, 4] is that all users participating in pseudonym change have the *same* anonymity set. However, from an individual user's perspective, the set of users that can obfuscate its pseudonym (i.e., its anonymity set) can be different from that of another user, depending on users' physical locations. For example, a user with a higher level of privacy sensitivity can have a smaller anonymity set than others. It is thus desirable to meet users' needs for *personalized location privacy*. To this end, we consider a general anonymity model where a user can define its specific anonymity set different from others' (as illustrated in Fig. 5.1).

In this chapter, we leverage the social tie structure among mobile users to incentivize them to participate in pseudonym change. To this end, we cast users' decision making of whether to participate in pseudonym change as a SGUM game, based on a general anonymity model that allows each user to have its specific anonymity set.

X. Gong et al., *Social Group Utility Maximization*, SpringerBriefs in Electrical and Computer Engineering, DOI 10.1007/978-3-319-12322-6_5

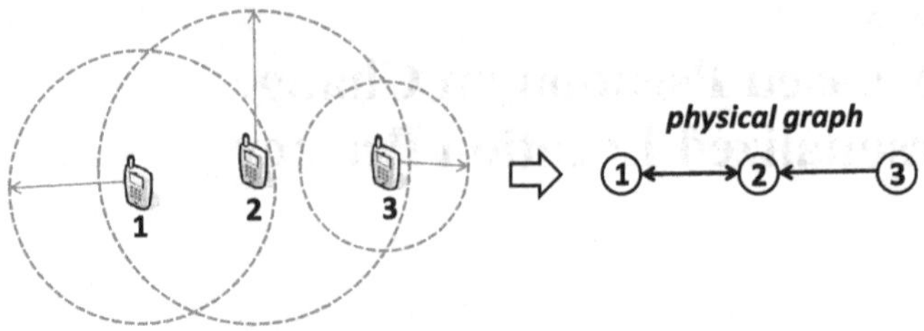

Fig. 5.1 Illustration of a general anonymity model: each user specifies its anonymity set for personalized location privacy by defining an anonymity range, e.g., a disk centered at the user's location. User 1 and 2 are out of user 3's anonymity range and thus are not in user 3's anonymity set (represented by no direct edge from user 1 or 2 to user 3); user 1 and 3 are within user 2's anonymity range and thus are in user 2's anonymity set (represented by directed edges from user 1 and 3 to user 2)

5.2 System Model

We consider a mobile network where users obtain their locations via mobile devicesthat are capable of localization (e.g., by GPS or wireless access points based localization). Users send their locations to a LBS provider for a certain LBS (e.g., location-based navigation or recommendation), and the LBS provider feedbacks the desired results to the users based on their reported locations. To protect privacy, each user uses a pseudonym as its identity for the LBS.

As in [1, 4, 5], we assume that the LBS provider is untrusted, i.e., it may leak users' location traces to an adversary. For example, the adversary may steal the location data by hacking into the LBS system. The adversary aims to learn the real identity of a user by linking and analyzing the locations visited by the user's pseudonym. We also assume that users are honest-but-curious such that each user honestly follows the protocols with others (which will be discussed in Sect. 5.5.3), but is curious about others' private information. We further assume that the adversary may collude with a limited number of users to gain useful information for inferring a user's real identity.

The use of pseudonym allows short-term reference to a user (e.g., one pseudonym can be used for the navigation of an entire trip between two locations), which is useful for many LBSs and does not disclose private information. However, long-term linking among a user's locations should be prevented, as it may reveal sufficient information for inferring the user's real identity [6, 7, 8]. Although a user may hide explicit linking among its locations by changing its pseudonym, the adversary can still link different pseudonyms of the user by exploiting spatial-temporal correlation in its locations. For example, consider a user that visits location l_1 with pseudonym Alice at time t_1, and then visits location l_2 that is close to location l_1 with pseudonym Bob at time t_2. If the adversary observes from the location traces that no other user changes its pseudonym between time t_1 and t_2, or there exists such a user but it does not visit any location close to location l_1 or l_2, then the adversary can infer that pseudonym Alice and Bob must refer to the same user, since only the same user can visit both location l_1 and l_2 within the limited period between time t_1 and t_2.

To protect location privacy from inference attacks, an effective approach is based on the notion of anonymity: users in physical proximity can coordinate their pseudonym changes to happen simultaneously [1], so that the adversary cannot link their pseudonyms before the changes to their respective pseudonyms after the changes. Existing studies [2, 3, 4] assume that all users participating in pseudonym change have the *same* anonymity set. However, based on an individual user's belief of the adversary's power against its location privacy (e.g., the adversary's side information about that user), the set of users that it believes can obfuscate its pseudonym (i.e., its anonymity set) can be different from that of another user. Thus motivated, we consider a general anonymity model that can meet users' needs for personalized location privacy, depending on users' physical locations. In particular, each user specifies an *anonymity range* (a physical area) such that the set of users within the anonymity range constitute that user's *potential anonymity set*. For example, a user's anonymity range can be a disk centered at the user's location, with a large radius indicating a low level of privacy sensitivity (as illustrated in Fig. 5.1). Note that for two users at different locations, their anonymity ranges are different even when they have the same shape (e.g., two disks with the same radius but different centers), and thus their potential anonymity sets can be different.

Formally, consider a set of users $\mathcal{N} \triangleq \{1, \cdots, N\}$ where each user i makes a decision a_i on whether or not to participate in pseudonym change, denoted by $a_i = 1$ and $a_i = 0$, respectively. Based on users' physical locations, the privacy gain perceived by a user participating in pseudonym change depends on which users also participate. Each user i incurs a cost of $c_i > 0$ to participate in pseudonym change. This cost is due to a number of factors, e.g., the participating users should stop using the LBS for a period of time. Based on the general anonymity model, the physical coupling among users can be captured by a physical graph $(\mathcal{N}, \mathcal{E}^P)$, where user j is connected by a directed edge $e_{ji}^P \in \mathcal{E}^P$ to user i if user j is in user i's potential anonymity set, denoted by $\mathcal{N}_{i-}^P$ (i.e., $j \in \mathcal{N}_{i-}^P$). Note that the physical coupling between two users can be asymmetric. The privacy gain perceived by a participating user i is defined as its *anonymity set size*, i.e., the number of participating users in $\mathcal{N}_{i-}^P$. Note that the anonymity set size is a widely adopted privacy metric[1] for anonymity-based approaches. For example, k-anonymity is used as the privacy metric in [4, 5], where a user achieves location privacy if its pseudonym cannot be distinguished among k users. Then the individual utility of user i, denoted by u_i, is given by

$$u_i(a_i, \boldsymbol{a}_{-i}) \triangleq a_i \left(\sum_{j \in \mathcal{N}_{i-}^P} a_j - c_i \right) \tag{5.1}$$

[1] Another privacy metric is the entropy of the adversary's uncertainty of a user's pseudonym. However, it is usually difficult to compute since it requires probability distribution which is difficult to attain.

where $\boldsymbol{a}_{-i}$ denotes the vector of the strategies of all users except user i. If a user participates, its individual utility is its privacy gain minus its participation cost; otherwise, it is zero. Note that c_i is a relative cost compared to privacy gain.

To take into account the social ties among users, each user i aims to maximize its social group utility, defined as

$$f_i(a_i, \boldsymbol{a}_{-i}) \triangleq u_i(a_i, \boldsymbol{a}_{-i}) + \sum_{j \in \mathcal{N}_{i+}^S} s_{ij} u_j(a_j, \boldsymbol{a}_{-j}). \tag{5.2}$$

Note that a user does not need to know the individual utilities of its social neighbors (which may be their private information) to make the decision (as will be shown in Eq. (5.3)). In Sect. 5.5.3, we will discuss how social information can be used while preserving the privacy of users' real identities with each other.

Under the SGUM framework, users' socially-aware decision making for pseudonym change boils down to a social group utility maximization game. Specifically, each user $i \in \mathcal{N}$ is a *player* and its *strategy*[2] is $a_i \in \{0, 1\}$. Let $\boldsymbol{a} = (a_1, \cdots, a_n)$ denote the *strategy profile* consisting of all users' strategies. The *payoff* function of a user is defined as its social group utility function. Given the strategies of other users, each user i aims to choose the *best response* strategy that maximizes its social group utility:

$$\underset{a_i}{\text{maximize}} \;\; f_i(a_i, \boldsymbol{a}_{-i}), \; \forall i \in \mathcal{N}.$$

For the sake of system efficiency, a natural objective is to maximize the *social welfare* of the system, which is the total individual utility of all users denoted by $v(\boldsymbol{a}) \triangleq \sum_{i \in \mathcal{N}} u_i(\boldsymbol{a})$. A strategy profile $\boldsymbol{a}^* = (a_1^*, \cdots, a_n^*)$ is *social optimal* [9] if it achieves the maximum social welfare among all profiles, i.e., $v(\boldsymbol{a}^*) \geq v(\boldsymbol{a})$, $\forall \boldsymbol{a}$. Although the social optimal profile is the best outcome in terms of system efficiency, it is often not acceptable by all users. Then, it is desirable to achieve the "best" SNE, i.e., the SNE that achieves the maximum social welfare among all SNEs. For brevity, we will refer to this SNE as the best SNE.

Another desirable notion for system efficiency is *Pareto-optimality*. A strategy profile $\boldsymbol{a}^{po} = (a_1^{po}, \cdots, a_n^{po})$ is Pareto-optimal [9] if there does not exist a Pareto-superior profile $\boldsymbol{a}' = (a_1', \cdots, a_n')$ such that no user achieves a worse individual utility while at least one user achieves a better individual utility, i.e.,

$$u_i(a_i', \boldsymbol{a}_{-i}') \geq u_i(a_i^{po}, \boldsymbol{a}_{-i}^{po}), \; \forall i \in \mathcal{N}$$

with at least one strict inequality.

For the SGUM-based PCG, we are interested in answering the following important questions: Does the game admit a SNE? How can we efficiently find a SNE with desirable system efficiency?

---

[2] As we focus on pure strategies in this work, we use "strategy" and "action" interchangeably.

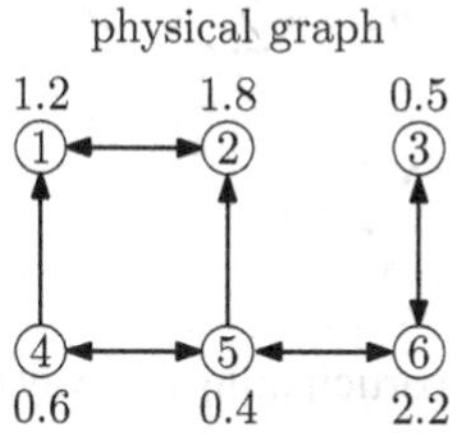

Fig. 5.2 Example of SO-PCG. The number beside a user is its cost. Using best response dynamics, we have $u_6 = 2 - 2.2 < 0 \rightarrow a_6 = 0 \rightarrow u_3 = 0 - 0.5 < 0 \rightarrow a_3 = 0 \rightarrow \mathcal{N}_1 = \{1, 2, 4, 5\}$, which is a SNE. It is also Pareto-superior to the other two SNEs: $\mathcal{N}_1 = \{4, 5\}$ and $\mathcal{N}_1 = \varnothing$, and hence is the best SNE

5.3 Benchmark: Socially-Oblivious Pseudonym Change Game

As the benchmark, we start with a basic case of the PCG: the PCG for socially-oblivious users (SO-PCG), i.e., $s_{ij} = 0, \forall e_{ij}^S \in \mathcal{N}^S$. In this case, each user is selfish and the social group utility degenerates to the individual utility.

For SO-PCG, there can exist multiple SNEs[3] with different values of social welfare (as illustrated in Fig. 5.2). For system efficiency, it is desirable to achieve the "best" SNE, which is the SNE that achieves the maximum social welfare among all SNEs. To find this SNE, we can use best response dynamics as follows: with all users' actions initially set to 1, each user asynchronously updates its action as its best response action based on other users' actions (no two users update at the same time). We illustrate how it works by an example in Fig. 5.2. We use $\mathcal{N}_1(\boldsymbol{a}) \triangleq \{i \in \mathcal{N} | a_i = 1\}$ to denote the set of participating users. The next result formally states that best response dynamics can find the best SNE.

Proposition 5.1 *For SO-PCG, best response dynamics can converge to a SNE that achieves the maximum social welfare among all SNEs.*

The proof is given in Appendix. As the best SNE achieves the maximum system efficiency among all SNEs, we will use the best SNE for SO-PCG as a benchmark for the general case of the PCG: the PCG for socially-aware users (SA-PCG).

5.4 Existence of SNE

We first establish the existence of SNE. Using (5.1) and (5.2), we have

$$f_i(1, \boldsymbol{a}_{-i}) - f_i(0, \boldsymbol{a}_{-i})$$

[3] For SO-PCG, an SNE is equivalent to a NE for a standard non-cooperative game. For consistency of terminology, we still call it "SNE" in this case.

$$= u_i(1, \boldsymbol{a}_{-i}) - u_i(0, \boldsymbol{a}_{-i}) + \sum_{j \in \mathcal{N}_{i+}^S} s_{ij} \left(u_j(1, \boldsymbol{a}_{-i}) - u_j(0, \boldsymbol{a}_{-i}) \right)$$

$$= \sum_{j \in \mathcal{N}_{i-}^P} a_j - c_i + \sum_{j \in \mathcal{N}_{i+}^S} s_{ij} a_j. \tag{5.3}$$

It is clear from (5.3) that no user participating is always a SNE. We therefore conclude that at least one SNE exists.

Then we show an important property of the social group utility function. It follows from (5.3) that

$$f_i(1, \boldsymbol{a}_{-i}) - f_i(0, \boldsymbol{a}_{-i}) - \left(f_i(1, \boldsymbol{a}'_{-i}) - f_i(0, \boldsymbol{a}'_{-i}) \right)$$

$$= \sum_{j \in \mathcal{N}_{i-}^P} a_j - c_i + \sum_{j \in \mathcal{N}_{i+}^S} s_{ij} a_j - \left(\sum_{j \in \mathcal{N}_{i-}^P} a'_j - c_i + \sum_{j \in \mathcal{N}_{i+}^S} s_{ij} a'_j \right)$$

$$= \sum_{j \in \mathcal{N}_{i-}^P} (a_j - a'_j) + \sum_{j \in \mathcal{N}_{i+}^S} s_{ij} (a_j - a'_j) \tag{5.4}$$

Let $\boldsymbol{a} \leq \boldsymbol{a}'$ denote element-wise inequality (i.e., $a_i \leq a'_i$, $\forall i \in \mathcal{N}$). The property below follows from (5.4).

Property 5.1 (Supermodularity) If $\boldsymbol{a}_{-i} \leq \boldsymbol{a}'_{-i}$, then $f_i(1, \boldsymbol{a}_{-i}) - f_i(0, \boldsymbol{a}_{-i}) \leq f_i(1, \boldsymbol{a}'_{-i}) - f_i(0, \boldsymbol{a}'_{-i})$.

Property 5.1 implies that if a user's best response strategy is to participate, then it remains the best response strategy if more users participate; if a user's best response strategy is to not participate, then it remains the best response strategy if less users participate.

5.5 Computing Pareto-Optimal SNE

Next we turn our attention to finding a SNE with desirable system efficiency.

For the PCG for fully altruistic users (i.e., $s_{ij} = 1$, $\forall e_{ij}^S \in \mathcal{N}^S$), it is clear that the social optimal profile $\boldsymbol{a}^*$ is a SNE, which is the solution to the following problem:

$$\begin{aligned} \underset{\boldsymbol{a}}{\text{maximize}} \quad & \sum_{i \in \mathcal{N}} a_i \left(\sum_{j \in \mathcal{N}_{i-}^P} a_j - c_i \right) \\ \text{subject to} \quad & a_i \in \{0, 1\}, \ \forall i \in \mathcal{N}. \end{aligned} \tag{5.5}$$

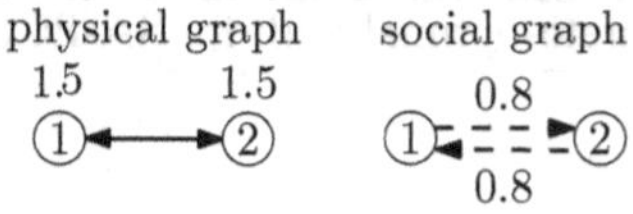

Fig. 5.3 Using best response dynamics, we have $f_1(1,1)-f_1(0,1)=1-1.5+0.8>0$, $f_2(1,1)-f_2(1,0)=1-1.5+0.8>0$, and hence $\mathcal{N}_1=\{1,2\}$ is a SNE. However, it is not Pareto-optimal, since it is Pareto inferior to $\mathcal{N}_1=\varnothing$ as $u_1(0,0)=u_2(0,0)=0>1-1.5=u_1(1,1)=u_2(1,1)$. Furthermore, its social welfare is less than that of $\mathcal{N}_1=\varnothing$ as $v(1,1)=-1<0=v(0,0)$, where $\mathcal{N}_1=\varnothing$ is also a SNE for SO-PCG

Observe that problem (5.5) is an integer quadratic programming, which is difficult to solve in general[4]. Since the PCG for fully altruistic users is a special case of SA-PCG, it is also difficult to compute the best SNE for SA-PCG. Based on this observation, our objective below is to efficiently compute a SNE with desirable system efficiency.

To compute a SNE of SA-PCG, a naive approach is to use best response dynamics in a similar way as with SO-PCG: with all users' actions initially set to 1, each user asynchronously updates its action as its best response action based on other users' actions. Due to Property 5.1, a user who changes its strategy from 1 to 0 will never change it back to 1, and thus the best response dynamics always converges to a SNE. However, it has drawbacks as illustrated by an example in Fig. 5.3: the SNE may not be Pareto-optimal and its social welfare may be worse than that of a SNE for SO-PCG. Thus motivated, our objective below is to efficiently find a SNE such that (1) it is Pareto-optimal and (2) its social welfare is no less than that of the best SNE for SO-PCG, which is the benchmark.

5.5.1 Algorithm Design

To this end, we design an algorithm as described in Algorithm 2. The main idea of the algorithm is to greedily determine users' strategies, depending on the social group utility derived from *the users whose strategies have been determined* (referred as "determined users"), denoted by

$$f_i'(a_i,\boldsymbol{a}_{-i}) \triangleq u_i(a_i,\boldsymbol{a}_{-i}) + \sum_{j\in\mathcal{N}_{i+}^S\setminus\overline{\mathcal{N}}} s_{ij}u_j(a_i,\boldsymbol{a}_{-i})$$

where $\overline{\mathcal{N}}$ denotes the set of users whose strategies have not been determined (referred as "undetermined users"). An undetermined user's action is fixed once it becomes determined.

Specifically, the algorithm proceeds in rounds and each round consists of phase I and phase II. In phase I, with all undetermined users' actions initially set to 1, an

[4] We conjecture that problem (5.5) is an NP-hard problem.

Algorithm 2 Compute a Pareto-optimal SNE for SA-PCG

$\overline{\mathcal{N}} \leftarrow \mathcal{N}$;
repeat
    `// Phase I;`
    $\boldsymbol{a} \leftarrow (1,\cdots,1)$, $\mathcal{N}_I \leftarrow \overline{\mathcal{N}}$;
    while $\exists i \in \overline{\mathcal{N}}$ such that $u_i(1,\boldsymbol{a}_{-i}) + \sum_{j\in\mathcal{N}_{i+}^S \setminus \overline{\mathcal{N}}} s_{ij}a_j < 0$ **do**
        $a_i \leftarrow 0$, $\mathcal{N}_I \leftarrow \mathcal{N}_I \setminus \{i\}$;
    end while
    `// Phase II;`
    $\overline{\mathcal{N}} \leftarrow \overline{\mathcal{N}} \setminus \mathcal{N}_I$, $\mathcal{N}_{II} \leftarrow \varnothing$;
    while $\exists i \in \overline{\mathcal{N}}$ such that $u_i(1,\boldsymbol{a}_{-i}) + \sum_{j\in\mathcal{N}_{i+}^S \setminus \overline{\mathcal{N}}} s_{ij}a_j \geq 0$ **do**
        $a_i \leftarrow 1$, $\overline{\mathcal{N}} \leftarrow \overline{\mathcal{N}} \setminus \{i\}$, $\mathcal{N}_{II} \leftarrow \mathcal{N}_{II} \cup \{i\}$;
    end while
until $\mathcal{N}_I \cup \mathcal{N}_{II} = \varnothing$;
return $\boldsymbol{a}^e \leftarrow \boldsymbol{a}$;

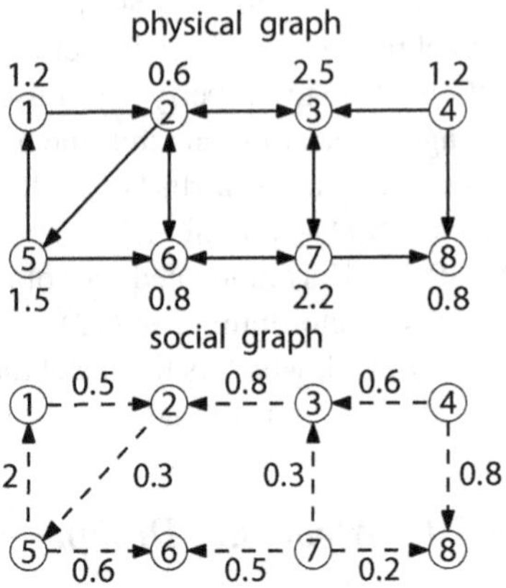

Fig. 5.4 Example to illustrate how Algorithm 2 works: The number next to a user is its cost; the number beside a social edge is its social tie level

undetermined user's action is changed from 1 to 0 if that improves its social group utility derived from the determined users, i.e.,

$$f_i'(1,\boldsymbol{a}_{-i}) - f_i'(0,\boldsymbol{a}_{-i}) = u_i(1,\boldsymbol{a}_{-i}) + \sum_{j\in\mathcal{N}_{i+}^S \setminus \overline{\mathcal{N}}} s_{ij}a_j < 0$$

until no such user exists. Then the undetermined users whose actions remain 1 become determined and their actions are fixed to 1. In phase II, with all undetermined users' actions initially set to 0, an undetermined user becomes determined and its action is fixed to 1 if that improves its social group utility derived from the determined users, until no such user exists. The algorithm terminates when no undetermined user becomes determined during either phase I or phase II of a round.

We use the example in Fig. 5.4 to illustrate how to compute a SNE using Algorithm 2 and outline the steps as follows.

- Phase I of 1st round: $u_1 = 1 - 1.2 < 0 \rightarrow a_1 = 0$; $u_5 = 1 - 1.5 < 0 \rightarrow a_5 = 0$; $u_4 = 0 - 1.2 < 0 \rightarrow a_4 = 0 \rightarrow u_3 = 2 - 2.5 < 0 \rightarrow a_3 = 0$; $u_7 = 2 - 2.2 < 0 \rightarrow a_7 = 0 \rightarrow u_8 = 0 - 0.8 < 0 \rightarrow a_8 = 0$; $u_2 = 1 - 0.6 > 0$; $u_6 = 1 - 0.8 > 0$; $\mathcal{N}_I = \{2, 6\}$.
- Phase II of 1st round: $u_5 + s_{56} = 1 - 1.5 + 0.6 > 0 \rightarrow a_5 = 1 \rightarrow u_1 + s_{12} = 1 - 1.2 + 0.5 > 0 \rightarrow a_1 = 1$; $u_3 + s_{32} = 1 - 2.5 + 0.8 < 0$; $u_7 + s_{76} = 1 - 2.2 + 0.5 < 0$; $u_4 = 0 - 1.2 < 0$; $u_8 = 0 - 0.8 < 0$; $\mathcal{N}_{II} = \{1, 5\}$.
- Phase I of 2nd round: $u_4 = 0 - 1.2 < 0 \rightarrow a_4 = 0 \rightarrow u_3 + s_{32} = 2 - 2.5 + 0.8 > 0$; $u_7 + s_{76} = 2 - 2.2 + 0.5 > 0$; $u_8 = 1 - 0.8 > 0$; $\mathcal{N}_I = \{3, 7, 8\}$.
- Phase II of 2nd round: $u_4 + s_{43} + s_{48} = 0 - 1.2 + 0.6 + 0.8 > 0 \rightarrow a_4 = 1$; $\mathcal{N}_{II} = \{4\}$.

Since the size of the set of undetermined users $\overline{\mathcal{N}}$ is upper bounded by n, the computational complexity of either phase I or phase II of a round is bounded by $O(n^2)$. Since at least one user is determined during a round, the algorithm must terminate within n rounds. Therefore, the running time of the algorithm is bounded by $O(n^3)$. In Sect. 5.6, numerical results will demonstrate that the computational complexity of Algorithm 2 increases almost quadratically with the number of users. In Sect. 5.5.2, we will discuss a distributed version of Algorithm 2.

Theorem 5.1 *For SA-PCG, the strategy profile* $\boldsymbol{a}^e = (a_1^e, \cdots, a_n^e)$ *computed by Algorithm 2 is a Pareto-optimal SNE.*

The proof is given in Appendix. As the SNE computed by Algorithm 2 is Pareto-optimal, it is desirable for system efficiency. Next we show that its social welfare is no less than that of the best SNE for SO-PCG. To this end, we first show that the Pareto-optimal SNE is monotonically "increasing" with respect to social tie levels.

Theorem 5.2 *For SA-PCG, when social tie levels increase (i.e.,* $s'_{ij} \geq s_{ij}, \forall i, j \in \mathcal{N}$*), the corresponding Pareto-optimal SNE* $\boldsymbol{a}^{e'}$ *satisfies that* $\boldsymbol{a}^{e'} \geq \boldsymbol{a}^e$ *and* $v(\boldsymbol{a}^{e'}) \geq v(\boldsymbol{a}^e)$.

The proof is given in Appendix. Intuitively, with stronger social ties to other users, a user is more likely to participate in favor of its social group utility, even at the cost of reducing its individual utility. Theorem 5.2 confirms this intuition: as social ties get stronger, more users participate at the Pareto-optimal SNE. Furthermore, the social welfare achieved at the Pareto-optimal SNE is also increasing.

When Algorithm 2 is used for SO-PCG, we can see that it works exactly the same as the best response dynamics used to find the best SNE for SO-PCG, and therefore they find the same profile. Based on this observation, using Theorem 5.2, we have the following result.

Corollary 5.1 *The social welfare of the Pareto-optimal SNE for SA-PCG is no less than that of the best SNE for SO-PCG.*

Corollary 5.1 guarantees that the social welfare of the Pareto-optimal SNE is no less than the benchmark SNE for SO-PCG. In Sect. 5.6, numerical results will demonstrate that the Pareto-optimal SNE is efficient, with a performance gain up to 20 % over the benchmark.

5.5.2 Distributed Computation of Pareto-optimal SNE

The Pareto-optimal SNE computed by Algorithm 2 can be achieved in a distributed manner. To this end, each user first obtains its potential anonymity set and its social tie levels with others. Following Algorithm 2, each user checks if it should change its strategy according to the condition in line 5 or 10 based on other users' strategies, and if yes, announces the change to all users. With time divided into slots, a random backoff mechanism can be used so that at most one user announces a change of strategy in a time slot. If no user announces a change, it indicates the end of phase I or phase II in Algorithm 2. Therefore, all users keep aware of the current state of the algorithm as it proceeds, and thus can act correctly according to the algorithm. The computational complexity of the distributed version of Algorithm 2 is almost the same as the centralized version, and is upper bounded by $O(n^3)$. Note that each user only knows the strategies of other users during the execution the algorithm, and thus users' privacy is preserved. After reaching the Pareto-optimal SNE, the users who decide to change their pseudonyms coordinate their pseudonym changes.

5.5.3 Further Discussions

We assume that there is a third party platform where users interact with each other to make pseudonym change decisions and coordinate their pseudonym changes. The platform only serves to allow information exchanges among users (e.g., an online chat service [10, 11]). We assume that the platform is honest-but-curious such that it honestly delivers messages among users, but is curious about users' private information. To protect privacy, each user also uses a pseudonym as its identity on the platform (which can be different from that used for the LBS). To make a socially-aware pseudonym change decision, each users needs to know its potential anonymity set and its social tie levels with others. This can be achieved in a privacy-preserving manner using secure protocols as discussed below. Note that the platform is not involved in the computing tasks of these protocols.

A user can learn whether another user is within its anonymity range using a certain private proximity detection protocol [12, 13]. For example, the protocol proposed in [12] can be used if the anonymity range is a disk. Specifically, the protocol involves several message exchanges between the two users, including one message that contains encrypted values that are functions of a user' location or the radius of the anonymity range. The protocol guarantees that both users can only learn the binary result of whether or not one is in another's anonymity range, and neither user can learn the other's location or anonymity range. In addition, since location information is encrypted in the messages, the platform cannot learn any user's location information. Similarly, the protocol in [13] can be used if the anonymity range is a convex polygon. Therefore, each user can learn its potential anonymity set without revealing its location information.

A user can also learn its social tie level with another user without disclosing one's real identity to the other. To this end, each user keeps a social profile consisting of the social communities that it belongs to (e.g., a community of colleagues at the same workplace), and sets a *single* social tie level for each community based on its social relationships with those in the community. Each community is identified by a predefined key that is only known to the community's members. Using a certain private matching protocol such as [14, 15], two users can learn whether they have a community in common, and if yes, which community[5] it is. In particular, the protocol involves several message exchanges between the two users, including one message that contains encrypted values that are functions of the keys of a user's social communities. The protocol ensures that both users can only know the community they have in common (if it exists), and neither user can learn any additional social information of the other, or pretend to have a community in common with the other. Since a community typically has many members, neither user can know the other's real identity even when they know the community they both belong to. In addition, since social information is encrypted in the messages, the platform cannot learn any user's social information. Therefore, each user can learn its social tie levels with those in its potential anonymity set while keeping their real identities private. Note that although the adversary might collude with multiple users, it is almost infeasible for the adversary to find a significant number of colluding users who have social ties with a specific user, in order to infer the user's real identity.

5.6 Numerical Results

In this section, we provide numerical results to illustrate the system efficiency of the Pareto-optimal SNE computed by Algorithm 2. We compare the social welfare of the Pareto-optimal SNE with the maximum social welfare of all SNEs for SO-PCG and the optimal social welfare, which is found by exhaustive search.

We consider N mobile users who are interested in participating in pseudonym change. They are randomly located in a square area with side length 500 m. We assume that the anonymity range of each user is a disk centered at the user's location with radius randomly chosen from $\{200\text{ m}, 300\text{ m}, 400\text{ m}\}$. Based on users' physical locations and anonymity ranges, there exists a physical edge from user i to user j if user i is in the anonymity range of user j. We assume that a user's cost of changing pseudonym is uniformly distributed in $[0, \overline{C}]$. We simulate the social graph using two methods as follows.

[5] To protect privacy, only one community in common is revealed if they have multiple communities in common.

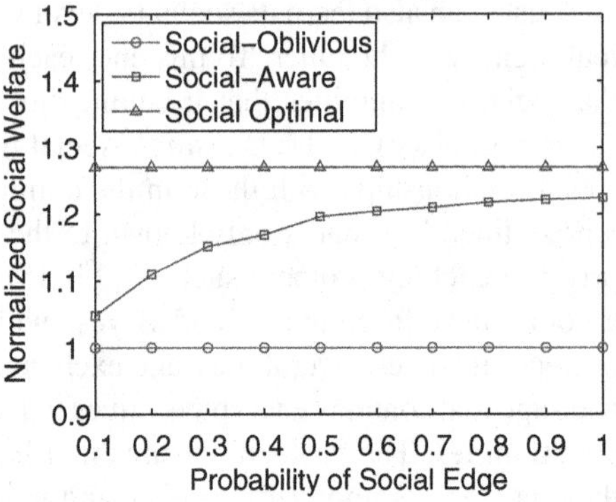

Fig. 5.5 Impact of P_S

5.6.1 Erdos-Renyi Model Based Social Graph

We simulate the social graph based on the Erdos-Renyi (ER) graph model [16], where a social edge exists between any two users with probability P_S. We assume that the level of a social tie is 1 if it exists. We set the default values of parameters as follows: $N = 10$, $P_S = 0.5$, $\overline{C} = 3$. For each set of parameter values, we average the results over 1000 runs.

We illustrate the impact of P_S, $\overline{C}$, and N on the normalized social welfare in Figs. 5.5, 5.6, and 5.7, respectively. We observe from these figures that socially-aware users significantly outperforms socially-oblivious users, especially when P_S or $\overline{C}$ is large, or N is small. This is due to that more users participate in pseudonym change when they are socially-aware, which improves the social welfare. On the other hand, the social welfare of socially-aware users is close to the optimal social welfare. Figure 5.5 shows that the performance of socially-aware users improves when P_S increases, with a performance gain up to 20 % over socially-oblivious users and a performance gap about 10 % on average from the optimal social welfare. This is due to that stronger social ties further encourage users to participate. Figures 5.6 and 5.7 show that the performance gap of socially-oblivious users compared to socially-aware users and the optimal social welfare, respectively, decreases as $\overline{C}$ or N increases. This is because that, with a lower participation cost or higher privacy gain of participation due to the increase of total user number, more users would participate even when they are socially-oblivious. Therefore, the performance gap reduces as it depends on the users that participate only when they are socially-aware. We plot the number of iterations for running Algorithm 2 versus N in Fig. 5.8. We observe that the computational complexity increases almost quadratically as the user number increases. This shows that our algorithm is scalable for a large number of users.

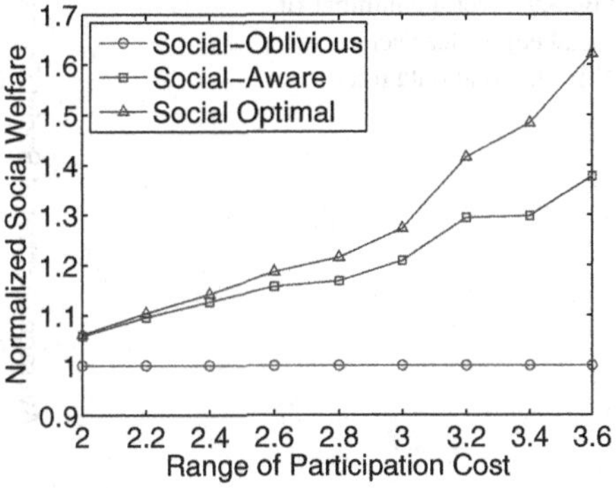

Fig. 5.6 Impact of $\overline{C}$

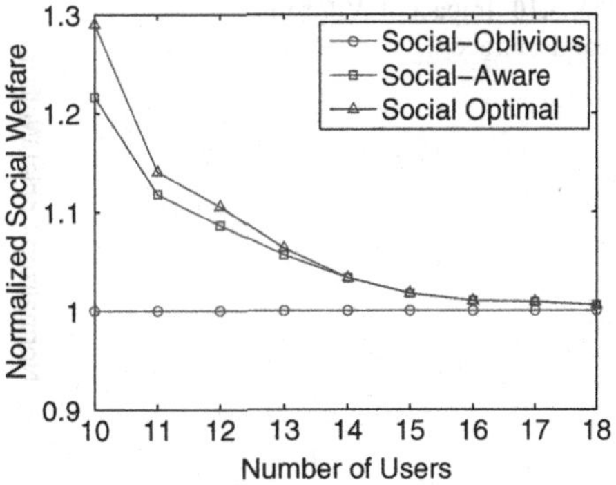

Fig. 5.7 Impact of N

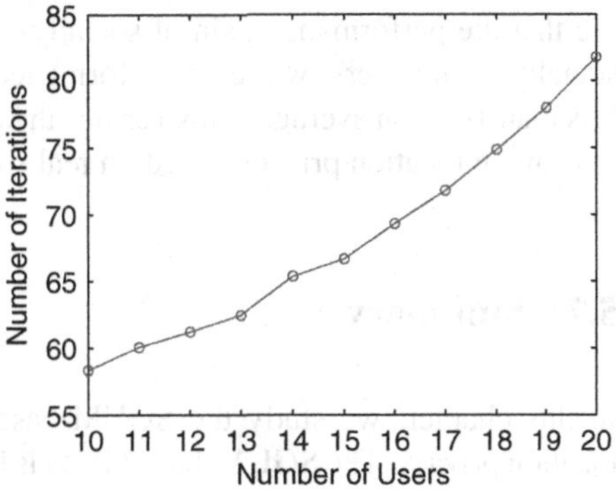

Fig. 5.8 Computational complexity versus N

5.6.2 *Real Data Trace Based Social Graph*

We simulate the social graph according to the social friendship network of the real data trace from Brightkite [17]. We plot the average number of social edges of a user versus the number of users in Fig. 5.9. We illustrate the impact of N on the social welfare in Fig. 5.10, where we set the range of participation cost as $\overline{C} = 3$. We can

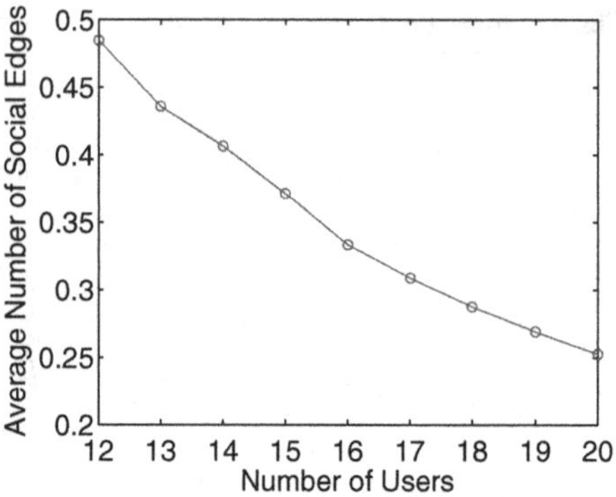

Fig. 5.9 Average number of social edges per user versus N for the real data trace

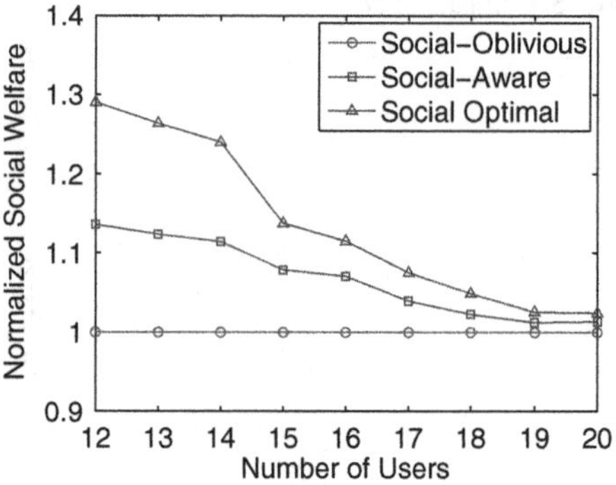

Fig. 5.10 Impact of N for the real data trace

see that the performance gain of socially-aware users can achieve up to 15 % over socially-aware users, while its performance gap from the optimal social welfare is less than 10 % on average. This verifies the effectiveness of exploiting social ties for improving location privacy based on real social data.

5.7 Summary

In this chapter, we study the SGUM-based pseudonym change for personalized location privacy. The SGUM-based PCG is based on a general anonymity model that allows each user to have its specific anonymity set. For the SGUM-based PCG, we show that there exists a SNE. Then we develop an algorithm that greedily determines users' strategies, based on the social group utility derived from only the users whose strategies have been determined. We show that this algorithm can efficiently find a Pareto-optimal SNE with social welfare higher than that corresponding to the socially-oblivious PCG. Numerical results demonstrate that social welfare can be significantly improved by exploiting users' social ties.

Appendix

For convenience, let $\mathcal{N}_{I,k}$ and $\mathcal{N}_{II,k}$ denote the set of users that become determined in phase I and phase II of round k in Algorithm 2, respectively. Let $\mathcal{N}_0(\boldsymbol{a}) \triangleq \{i \in \mathcal{N} | a_i = 0\}$ denote the set of users with action 0 under profile $\boldsymbol{a}$.

Proof of Proposition 5.1

Since SO-PCG is a special case of SA-PCG, Property 5.1 also applies to the individual utility function u_i. Therefore, due to Property 5.1, a user who changes its strategy from 1 to 0 will not change it back to 1. As a result, best response dynamics always terminates and results in a profile $\boldsymbol{a}^o$ that is a SNE.

Next we show that $\boldsymbol{a}^o$ achieves the maximum social welfare among all SNEs. It suffices to show that $\boldsymbol{a}^o$ is Pareto-superior to any other SNE. To this end, we first show that a profile $\boldsymbol{a}'$ is not a SNE if $\mathcal{N}_1(\boldsymbol{a}') \setminus \mathcal{N}_1(\boldsymbol{a}^o) \neq \varnothing$. Suppose such $\boldsymbol{a}'$ is a SNE. Let $i \in \mathcal{N}_1(\boldsymbol{a}') \setminus \mathcal{N}_1(\boldsymbol{a}^o)$ be the first user among $\mathcal{N}_1(\boldsymbol{a}') \setminus \mathcal{N}_1(\boldsymbol{a}^o)$ whose action is changed to 0, and $\bar{\boldsymbol{a}}$ be the profile right before that change. Since $\boldsymbol{a}' \leq \bar{\boldsymbol{a}}$, we have $u_i(\boldsymbol{a}') \leq u_i(\bar{\boldsymbol{a}}) < 0 = u_i(0, \boldsymbol{a}'_{i-})$ due to that 0 is the best response strategy. This shows that $\boldsymbol{a}'$ is not a SNE. Therefore, for any SNE $\boldsymbol{a}'$ other than $\boldsymbol{a}^o$, we must have $\boldsymbol{a}' < \boldsymbol{a}^o$. Then for each $i \in \mathcal{N}_1(\boldsymbol{a}')$, we have $u_i(\boldsymbol{a}') \leq u_i(\boldsymbol{a}^o)$. For each $i \in \mathcal{N}_0(\boldsymbol{a}')$, since $\boldsymbol{a}^o$ is a SNE, we have $u_i(\boldsymbol{a}') = 0 = u_i(0, \boldsymbol{a}^o_{-i}) \leq u_i(\boldsymbol{a}^o)$. Therefore $\boldsymbol{a}^o$ is Pareto-superior to $\boldsymbol{a}'$. Thus we show that $\boldsymbol{a}^o$ is the best SNE.

Proof of Theorem 5.1

We first show that $\boldsymbol{a}^e$ is a SNE. We consider three cases of a user i as follows.

Case 1: $i \in \mathcal{N}_1(\boldsymbol{a}^e)$ and $i \in \mathcal{N}_{I,k}$

Let $\boldsymbol{a}'$ be the profile right after phase I during which i remains in $\mathcal{N}_I$. Since $\boldsymbol{a}^e \geq \boldsymbol{a}'$, using (5.3) we have

$$f_i(1, \boldsymbol{a}^e_{-i}) - f_i(0, \boldsymbol{a}^e_{-i}) \geq u_i(1, \boldsymbol{a}'_{-i}) + \sum_{j \in \mathcal{N}^S_{i+} \setminus \overline{\mathcal{N}}} s_{ij} a'_j \geq 0$$

where the second inequality is due to the condition in line 5.

Case 2: $i \in \mathcal{N}_1(\boldsymbol{a}^e)$ and $i \in \mathcal{N}_{II,k}$

Let $\boldsymbol{a}'$ be the profile right after i becomes determined in phase II. Since $\boldsymbol{a}^e \geq \boldsymbol{a}'$, using (5.3) we have

$$f_i(1, \boldsymbol{a}^e_{-i}) - f_i(0, \boldsymbol{a}^e_{-i}) \geq u_i(1, \boldsymbol{a}'_{-i}) + \sum_{j \in \mathcal{N}^S_{i+} \setminus \overline{\mathcal{N}}} s_{ij} a'_j \geq 0$$

where the second inequality is due to the condition in line 10.

Case 3: $i \in \mathcal{N}_0(\boldsymbol{a}^e)$

Since i is not included in $\mathcal{N}_{II}$ in phase II of the last round, using (5.3) we have

$$f_i(1, \boldsymbol{a}^e_{-i}) - f_i(0, \boldsymbol{a}^e_{-i}) = u_i(1, \boldsymbol{a}'_{-i}) + \sum_{j \in \mathcal{N}^S_{i+} \setminus \overline{\mathcal{N}}} s_{ij} a'_j < 0$$

where the inequality is due to the condition in line 10.

Next we show that $\boldsymbol{a}^e$ is Pareto-optimal. Suppose there exists $\boldsymbol{a}'$ that is Pareto-superior to $\boldsymbol{a}^e$. It suffices to show that (i) $\mathcal{N}_1(\boldsymbol{a}') \setminus \mathcal{N}_1(\boldsymbol{a}^e) = \varnothing$ and (ii) $\mathcal{N}_1(\boldsymbol{a}^e) \setminus \mathcal{N}_1(\boldsymbol{a}') = \varnothing$. We first show part (i). Suppose $\mathcal{N}_1(\boldsymbol{a}') \setminus \mathcal{N}_1(\boldsymbol{a}^e) \neq \varnothing$. Then for each $i \in \mathcal{N}_1(\boldsymbol{a}') \setminus \mathcal{N}_1(\boldsymbol{a}^e)$, we have $u_i(\boldsymbol{a}') \geq u_i(\boldsymbol{a}^e) = 0$. Let i be the first user among $\mathcal{N}_1(\boldsymbol{a}') \setminus \mathcal{N}_1(\boldsymbol{a}^e)$ whose action is set to 0 during phase I of the last round, and $\bar{\boldsymbol{a}}$ be the profile right before $a_i = 0$ is performed. Since $\bar{\boldsymbol{a}}_{-i} \geq \boldsymbol{a}'_{-i}$, we have $u_i(1, \bar{\boldsymbol{a}}_{-i}) + \sum_{j \in \mathcal{N}^S_{i+} \setminus \overline{\mathcal{N}}} s_{ij} \bar{a}_j \geq u_i(1, \bar{\boldsymbol{a}}_{-i}) \geq u_i(1, \boldsymbol{a}'_{-i}) \geq 0$, which contradicts to the condition in line 5.

Next we show part (ii). Suppose $\mathcal{N}_1(\boldsymbol{a}^e) \setminus \mathcal{N}_1(\boldsymbol{a}') \neq \varnothing$. Since we have shown part i), we must have $\boldsymbol{a}' < \boldsymbol{a}^e$. Then for each $i \in \mathcal{N}_1(\boldsymbol{a}') \subset \mathcal{N}_1(\boldsymbol{a}^e)$, we have $u_i(\boldsymbol{a}') \leq u_i(\boldsymbol{a}^e)$. Since $u_i(\boldsymbol{a}') = 0 = u_i(\boldsymbol{a}^e)$ for each $i \in \mathcal{N}_0(\boldsymbol{a}') \cap \mathcal{N}_0(\boldsymbol{a}^e)$, there must exist $i \in \mathcal{N}_1(\boldsymbol{a}^e) \setminus \mathcal{N}_1(\boldsymbol{a}')$ such that $u_i(\boldsymbol{a}^e) < u_i(\boldsymbol{a}') = 0$. Suppose i is included in $\mathcal{N}_{II}$ during phase II of some round. Let $\bar{\boldsymbol{a}}$ be the profile right before $a_i = 1$ is performed. Since $\bar{\boldsymbol{a}} \leq \boldsymbol{a}^e$, we have $u_i(\bar{\boldsymbol{a}}) \leq u_i(\boldsymbol{a}^e) < 0$. Then it follows from $0 \leq f_i(1, \bar{\boldsymbol{a}}_{-i}) - f_i(0, \bar{\boldsymbol{a}}_{-i}) = u_i(1, \bar{\boldsymbol{a}}_{-i}) + \sum_{j \in \mathcal{N}^S_{i+}} s_{ij} \bar{a}_j$ that there must exist $j \in \mathcal{N}^S_{i+}$ such that $\bar{a}_j = 1$, and therefore $a^e_j = 1$. If $j \in \mathcal{N}_1(\boldsymbol{a}')$, we have $u_j(\boldsymbol{a}^e) - u_j(\boldsymbol{a}') \geq a^e_i - a'_i = 1 > 0$, which is a contradiction. Therefore we must have $j \in \mathcal{N}_1(\boldsymbol{a}^e) \setminus \mathcal{N}_1(\boldsymbol{a}')$ and $u_j(\boldsymbol{a}^e) \leq u_j(\boldsymbol{a}') = 0$. Let $\hat{\boldsymbol{a}}$ be the profile right before $a_j = 1$ is performed. Since j is included before i, we have $u_j(\hat{\boldsymbol{a}}) < u_j(\boldsymbol{a}^e) \leq 0$. Then we can use the above argument sequentially, until we find some k that leads to contradiction.

Proof of Theorem 5.2

Let $\mathcal{N}'_{I,k}$ be the set of users in $\mathcal{N}_{I,k}$ during the execution that computes $\boldsymbol{a}^{e'}$. For each $i \in \mathcal{N}_{I,1}$, we have

$$u_i(1, \boldsymbol{a}'_{-i}) + \sum_{j \in \mathcal{N}^{S'}_{i+} \setminus \overline{\mathcal{N}}'} s'_{ij} a_j \geq u_i(1, \boldsymbol{a}_{-i}) + \sum_{j \in \mathcal{N}^S_{i+} \setminus \overline{\mathcal{N}}} s_{ij} a_j \geq 0.$$

Therefore we must have $\mathcal{N}_{I,1} \subseteq \mathcal{N}'_{I,1}$. Similarly, we can show that for any $i \in \mathcal{N}_{II,1} \setminus \mathcal{N}'_{I,1}$, we must have $i \in \mathcal{N}'_{II,1}$. Using this argument sequentially, we can show that $\cup_{i=1}^k (\mathcal{N}_{I,i} \cup \mathcal{N}_{II,i}) \subseteq \cup_{i=1}^k (\mathcal{N}'_{I,i} \cup \mathcal{N}'_{II,i})$ for any k, and therefore $\boldsymbol{a}^e \leq \boldsymbol{a}^{e'}$. When a user becomes determined with action 1, the increment of social welfare of determined users by changing its action from 0 to 1 is no less than the increment of its social group utility derived from determined users, which is non-negative. Therefore we can see that $v(\boldsymbol{a}^e) \leq v(\boldsymbol{a}^{e'})$.

References

1. A. Beresford, F. Stajano, Location privacy in pervasive computing. IEEE Pervasive Comput. **2**(1), 46–55 (2003)
2. J. Freudiger, M.H. Manshaei, J.-P. Hubaux, D.C. Parkes, On non-cooperative location privacy: A game-theoretic analysis. ACM CCS. 324–337 (2009)
3. X. Liu, H. Zhao, M. Pan, H. Yue, X. Li, Y. Fang, Traffic-aware multiple mix zone placement for protecting location privacy. IEEE INFOCOM. 972–980 (2012)
4. D. Yang, X. Fang, G. Xue, Truthful incentive mechanisms for k-anonymity location privacy. IEEE INFOCOM. 2994–3002 (2013)
5. X. Liu, K. Liu, L. Guo, X. Li, Y. Fang, A game-theoretic approach for achieving k-anonymity in location based services. IEEE INFOCOM. 2985–2993 (2013)
6. B. Hoh, M. Gruteser, Protecting location privacy through path confusion. SECURECOMM. 194–205 (2005)
7. B. Hoh, M. Gruteser, H. Xiong, A. Alrabady, Enhancing security and privacy in traffic-monitoring systems. IEEE Pervasive Comput. 38–46 (2006)
8. J. Krumm, Inference attacks on location tracks. IEEE Pervasive Comput. 127–143 (2007)
9. M. Osborne, A. Rubinstein, *A Course in Game Theory* (MIT Press, Cambridge, 1994)
10. WeChat: The new way to connect (2014), http://www.wechat.com/
11. WhatsApp: Simple. Personal. Real time messaging (2014), http://www.whatsapp.com/
12. G. Zhong, I. Goldberg, U. Hengartner, Louis, Lester and Pierre: Three protocols for location privacy. ACM PETS. 62–76 (2007)
13. B. Mu, S. Bakiras, Private proximity detection for convex polygons. ACM MobiDE. 36–43 (2013)
14. R. Zhang, Y. Zhang, J. Sun, G. Yan, Fine-grained private matching for proximity-based mobile social networking. IEEE INFOCOM. 1969–1977 (2012)
15. L. Zhang, X.-Y. Li, Y. Liu, Message in a sealed bottle: Privacy preserving friending in social networks. IEEE ICDCS. 327–336 (2013)
16. P. Erdos, A. Renyi, On the evolution of random graphs. Publ. Math. Inst. Hung. Acad. Sci. 17–61 (1960)
17. SNAP: Network datasets: Brightkite (2014), http://snap.stanford.edu/data/loc-brightkite.html

References

Chapter 6
Conclusion and Future Work

6.1 Conclusion

With continuing technological advances, the past few years have witnessed the pervasive penetration of wireless networks in people's lives. On the other hand, social relationships play an increasingly important role in people's interactions with each other, due to the rapid growth of online social networking services. In this brief, we advocate to leverage wireless users' social ties to stimulate their cooperative behaviors in order to enhance their interactions in wireless networks. In general, wireless users have diverse social ties while their wireless devices have diverse physical relationships. To capture both the social coupling and physical coupling among wireless users, we develop a social group utility maximization (SGUM) framework. In the SGUM game, each user aims to maximize its social group utility, which is the sum of its own utility and the weighted sum of the utilities of the users having social tie with it. To illustrate how to apply the SGUM framework for cooperative wireless networking, we study its application in some specific contexts as follows.

- We study the SGUM-based random access control and power control. For the SGUM-based random access control game, we derive the unique SNE. For the SGUM-based power control game, we show that it is a supermodular game and thus there exists an SNE. We also derive the unique SNE for the two-user case of the SGUM-based power control game. For both games, we show that as social tie levels increase, each user's SNE strategy is decreasing and the social welfare of the SNE is increasing.
- We study the SGUM-based database assisted spectrum access. We show that the SGUM-based spectrum access game is a potential game and thus always admits a SNE. Then we design a distributed spectrum access algorithm that can achieve the socially-aware Nash equilibrium. We also derive the upper-bound of the performance gap of the socially-aware Nash equilibrium from the NUM solution. Numerical results demonstrate that the performance gap between the SGUM solution and the NUM solution is at most 15 %.
- We study the SGUM-based pseudonym change for personalized location privacy. The SGUM-based pseudonym change game (PCG) is based on a general

X. Gong et al., *Social Group Utility Maximization*, SpringerBriefs in Electrical and Computer Engineering, DOI 10.1007/978-3-319-12322-6_6

anonymity model that allows each user to have its specific anonymity set. For the SGUM-based PCG, we show that there exists a SNE. Then we develop an algorithm that greedily determines users' strategies, based on the social group utility derived from only the users whose strategies have been determined. We show that this algorithm can efficiently find a Pareto-optimal SNE with social welfare higher than that corresponding to the socially-oblivious PCG. Numerical results demonstrate that social welfare can be significantlyimproved by exploiting users' social ties.

6.2 Future Work

In Chap. 2, we develop the SGUM framework which leverage user's "positive" social ties to stimulate their cooperative behaviors. In general, depending on the nature of social relationship, the social tie between two users can be "negative" (e.g., between opponents or enemies) such that one user intends to damage the other's welfare. It is thus natural to extend the SGUM framework to capture negative social ties. Similar to positive social ties, negative social ties can also be diverse such that a user intends to damage others at different levels.

To capture both positive and "negative" social ties, we can extend the SGUM framework by defining the social group utility f_i as

$$f_i(x_i, \boldsymbol{x}_{-i}) \triangleq \sum_{j=1}^{N} s_{ij} u_i(x_i, \boldsymbol{x}_{-i})$$

where $s_{ij} \in (-\infty, 1]$: when $s_{ij} \in (0, 1]$, it represents the extent to which user i cares about user j's utility, and it reaches the maximum when $s_{ij} = 1$ (i.e., user i cares about user j's utility as much as its own utility); when $s_{ij} \in (-\infty, 0)$, it pinpoints to how much user i intends to damage user j's utility, and reaches the extreme as s_{ij} goes to $-\infty$ (i.e., user i would sacrifice its utility to damage user j's utility).

For the extended SGUM game, if the total social tie level to each user is 0, i.e.,

$$\sum_{j \in \mathcal{N}} s_{ji} = 0, \ \forall i \in \mathcal{N},$$

then the SGUM game degenerates to a zero-sum game (ZSG), where each user views the total gain of other users as its loss. Formally, a game is a ZSG if all users' payoff functions are summed up to 0. For example, an SGUM game of two users with $f_1 = u_1 - u_2$ and $f_2 = u_2 - u_1$, or $f_1 = u_1$ and $f_2 = -u_1$, is a zero-sum game. Note that we obtain an equivalent game when a user's payoff function is multiplied by a number or added by a function independent of that user's strategy. For example, an SGUM game of two users with $f_1 = u_1$ and $f_2 = u_2 - s_{21}u_1$ where $s_{21} \to \infty$ is equivalent to that with $f_1 = u_1$ and $f_2 = -u_1$, and thus is a zero-sum game. Therefore, the extended SGUM framework encompasses not only NCG and NUM but also ZSG as special cases (as illustrated in Fig. 6.1). Furthermore, it spans the continuum from ZSG to NCG to NUM (as illustrated in Fig. 6.2).

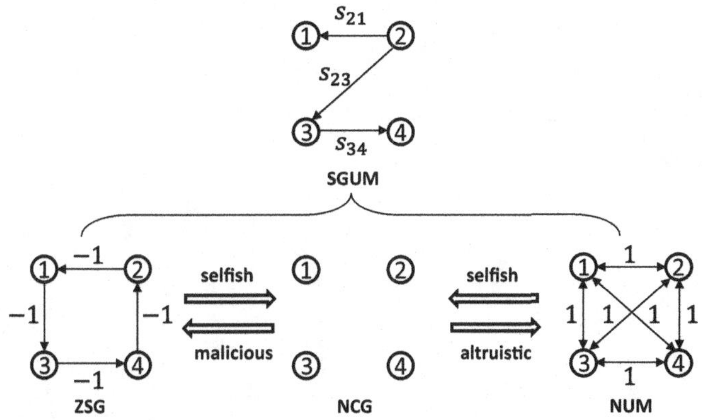

Fig. 6.1 The extended social group utility maximization (*SGUM*) game captures zero-sum game (*ZSG*), non-cooperative game (*NCG*), and network utility maximization (*NUM*) as special cases

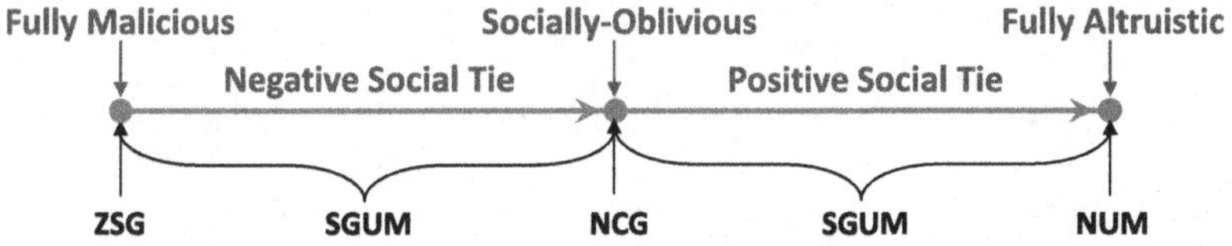

Fig. 6.2 The social group utility maximization (*SGUM*) game framework spans the continuum space from zero-sum game (*ZSG*) to non-cooperative game (*NCG*) to network utility maximization (*NUM*)

The integration of "negative" social ties under the SGUM framework offers a social perspective on network security: the "adversary" in the context of network security can be viewed as a malicious user who has "negative" social ties with other users. As the malicious user can have diverse "negative" social ties with other users, an interesting future direction is to investigate how the diverse social tie structure would impact the malicious user's behavior. Furthermore, the rich modeling flexibility provided by the extended SGUM framework allows us to study the tradeoff between security and utility. As a malicious user's attack would result in other users' utility loss, an important question is how it would depend on the malicious user's "negative" social tie levels with other users.